About the Authors

GARY SMALL, M.D., is the director of the Memory and Aging Research Center at the Semel Institute for Neuroscience and Human Behavior and the Center on Aging at UCLA. His research has made the headlines of the *Wall Street Journal*, the *New York Times*, and *USA Today*, among other publications. *Scientific American* magazine has named him one of the world's top innovators in science and technology.

GIGI VORGAN wrote, produced, and appeared in numerous feature films and television shows before teaming up with her husband, Dr. Gary Small, to co-write *The Memory Bible, The Memory Prescription, The Longevity Bible,* and *iBrain*. She lives in Los Angeles with Dr. Small and their two children.

Author photographs © Sterling Franken-Steffen

iBrain

Also by Gary Small, M.D.

The Memory Bible
with Gigi Vorgan

The Memory Prescription

The Longevity Bible

iBrain

Surviving the Technological Alteration of the Modern Mind

Gary Small, M.D.
and Gigi Vorgan

HARPER

NEW YORK · LONDON · TORONTO · SYDNEY

HARPER

To book Dr. Gary Small for a speaking engagement, visit: www.harpercollinsspeakers.com.

A hardcover edition of this book was published in 2008 by Collins Living, an imprint of HarperCollins Publishers.

HarperCollins books may be purchased for educational, business, or sales promotional use. For information please write: Special Markets Department, HarperCollins Publishers, 10 East 53rd Street, New York, NY 10022.

FIRST HARPER PAPERBACK PUBLISHED 2009.

Illustrations by Diana Jacobs
Designed by Level C

The Library of Congress has catalogued the hardcover edition as follows:

Small, Gary W.
  iBrain: surviving the technological alteration of the modern mind / by Gary Small and Gigi Vorgan. — 1st ed.
      p. cm.
  Includes bibliographical references and index.
  ISBN-13: 978-0-06-134033-8
  1. Brain—Evolution.   2. Digital media—Psychological aspects.   3. Neuroplasticity.
I. Vorgan, Gigi, 1958-   II. Title.
  QP376.S6377 2008
  612.8'20285—dc22   2007049587

ISBN 978-0-06-134034-5 (pbk.)
10 11 12 13 WBC/RRD 10 9 8 7 6 5 4 3 2

This book is dedicated to Rachel and Harry,
our own Digital Natives,
and all the *future brains* of the world.

ACKNOWLEDGMENTS

We wish to thank the many scientists and innovators whose work inspired this book, as well as our friends and colleagues who contributed their energy and insights, including Rachel Champeau, Kim Dower, Sterling Franken-Steffen, Stephanie Oudiz, Pauline Spaulding, and Cara and Rob Steinberg. We are also indebted to our talented artist and friend Diana Jacobs, for her creative drawings included in this book. We also appreciate the Parvin Foundation and Drs. Susan Bookheimer and Teena Moody for supporting and contributing to our new study, "Your Brain on Google."

iBrain would not have been possible without the support and input from our editor extraordinaire, Mary Ellen O'Neill, and our longtime agent and good friend, Sandra Dijkstra. We also want to thank our children, Rachel and Harry, as well as our parents, Dr. Max and Gertrude Small, and Rose Vorgan and Fred Weiss, for their love and encouragement.

Gary Small, M.D.
Gigi Vorgan

CONTENTS

Contents

iBrain

One

YOUR BRAIN IS
EVOLVING RIGHT NOW

> *The people who are crazy enough to think they*
> *can change the world are the ones who do.*
>
> Steve Jobs, CEO of Apple

You're on a plane packed with other business people, reading your electronic version of the *Wall Street Journal* on your laptop while downloading files to your BlackBerry and organizing your PowerPoint presentation for your first meeting when you reach New York. You relish the perfect symmetry of your schedule, to-do lists, and phone book as you notice a woman in the next row entering little written notes into her leather-bound daily planner book. You remember having one of those . . . What? Like a zillion years ago? Hey lady! Wake up and smell the computer age.

You're outside the airport now, waiting impatiently for a cab along with a hundred other people. It's finally your turn, and as you reach for the taxi door a large man pushes in front of you, practically knocking you over. Your briefcase goes flying, and your laptop and BlackBerry splatter into pieces on the pavement. As you frantically gather up the remnants of your once perfectly scheduled life, the woman with the daily planner book gracefully steps into a cab and glides away.

The current explosion of digital technology not only is changing the way we live and communicate but is rapidly and profoundly altering our brains. Daily exposure to high technology—computers, smart phones, video games, search engines like Google and Yahoo—stimulates brain cell alteration and neurotransmitter release, gradually strengthening new neural pathways in our brains while weakening old ones. Because of the current technological revolution, our brains are *evolving* right now—at a speed like never before.

Besides influencing how we think, digital technology is altering how we feel, how we behave, and the way in which our brains function. Although we are unaware of these changes in our neural circuitry or brain wiring, these alterations can become permanent with repetition. This evolutionary brain process has rapidly emerged over a *single* generation and may represent one of the most unexpected yet pivotal advances in human history. Perhaps not since Early Man first discovered how to use a tool has the human brain been affected so quickly and so dramatically.

Television had a fundamental impact on our lives in the past century, and today the average person's brain continues to have extensive daily exposure to TV. Scientists at the University of California, Berkeley, recently found that on average Americans spend nearly three hours each day watching television or movies, or much more time spent than on *all* leisure physical activities combined. But in the current digital environment, the Internet is replacing television as the prime source of brain stimulation. Seven out of ten American homes are wired for high-speed Internet. We rely on the Internet and digital technology for entertainment, political discussion, and even social reform as well as communication with friends and co-workers.

As the brain evolves and shifts its focus toward new technological skills, it drifts away from fundamental social skills, such as reading facial expressions during conversation or grasping the emotional context of a subtle gesture. A Stanford University study found that for every hour we spend on our computers, traditional face-to-face interaction time with other people drops by nearly thirty minutes. With the weakening of the brain's neural circuitry controlling human contact, our social interactions may become awkward, and we tend to misinterpret, and even miss subtle, nonverbal messages. Imagine how the continued slipping of social skills might affect an international summit meeting ten years from now when a misread facial cue or a misunderstood gesture could make the difference between escalating military conflict or peace.

The high-tech revolution is redefining not only how we communicate but how we reach and influence people, exert political and social change, and even glimpse into the private lives of co-workers, neighbors, celebrities, and politicians. An unknown innovator can become an overnight media magnet as news of his discovery speeds across the Internet. A cell phone video camera can capture a momentary misstep of a public figure, and in minutes it becomes the most downloaded video on YouTube. Internet social networks like MySpace and Facebook have exceeded a hundred million users, emerging as the new marketing giants of the digital age and dwarfing traditional outlets such as newspapers and magazines.

Young minds tend to be the most exposed, as well as the most sensitive, to the impact of digital technology. Today's young people in their teens and twenties, who have been dubbed Digital Natives, have never known a world without computers, twenty-four-hour TV news, Internet, and cell phones—with their video, music, cameras, and text messaging. Many of these Natives rarely enter a library, let alone look something up in a traditional encyclopedia; they use Google, Yahoo, and other online search engines. The neural networks in the brains of these Digital Natives differ dramatically from those of Digital Immigrants: people—including all baby boomers—who came to the digital/computer age as adults but whose basic brain wiring was laid down during a time when direct social interaction was the norm. The extent of their early technological communication and entertainment involved the radio, telephone, and TV.

As a consequence of this overwhelming and early high-tech stimulation of the Digital Native's brain, we are witnessing the beginning of a deeply divided *brain gap* between younger and older minds—in just *one* generation. What used to be simply a *generation gap* that separated young people's values, music, and habits from those of their parents has now become a huge divide resulting in two separate cultures. The brains of the younger generation are digitally hardwired from toddlerhood, often at the expense of neural circuitry that controls one-on-one people skills. Individuals of the older generation face a world in which their brains *must* adapt to high technology, or they'll be left behind—politically, socially, and economically.

Young people have created their own digital social networks, including a shorthand type of language for text messaging, and studies show that fewer young adults read books for pleasure now than in any generation before them. Since 1982, literary reading has declined by 28 percent in eighteen- to thirty-four-year-olds. Professor Thomas

Patterson and colleagues at Harvard University reported that only 16 percent of adults age eighteen to thirty read a daily newspaper, compared with 35 percent of those thirty-six and older. Patterson predicts that the future of news will be in the electronic digital media rather than the traditional print or television forms.

These young people are not abandoning the daily newspaper for a stroll in the woods to explore nature. Conservation biologist Oliver Pergams at the University of Illinois recently found a highly significant correlation between how much time people spend with new technology, such as video gaming, Internet surfing, and video watching, and the decline in per capita visits to national parks.

Digital Natives are snapping up the newest electronic gadgets and toys with glee and often putting them to use in the workplace. Their parents' generation of Digital Immigrants tends to step more reluctantly into the computer age, not because they don't want to make their lives more efficient through the Internet and portable devices but because these devices may feel unfamiliar and might upset their routine at first.

During this pivotal point in brain evolution, Natives and Immigrants alike can learn the tools they need to take charge of their lives and their brains, while both preserving their humanity and keeping up with the latest technology. We don't all have to become techno-zombies, nor do we need to trash our computers and go back to writing longhand. Instead, we all should help our brains adapt and succeed in this ever-accelerating technological environment.

IT'S ALL IN YOUR HEAD

Every time our brains are exposed to new sensory stimulation or information, they function like camera film when it is exposed to an image. The light from the image passes through the camera lens and causes a chemical reaction that alters the film and creates a photograph.

As you glance at your computer screen or read this book, light impulses from the screen or page will pass through the lens of your eye and trigger chemical and electrical reactions in your retina, the membrane in the back of the eye that receives images from the lens and sends them to the brain through the optic nerve. From the optic nerve, neurotransmitters send their messages through a complex network of neurons, axons, and dendrites until you become consciously aware of the screen or page. All this takes a miniscule fraction of a second.

Perception of the image may stir intense emotional reactions, jog repressed memories, or simply trigger an automatic physical response—like turning the page or scrolling down the computer screen. Our moment-to-moment responses to our environment lead to very particular chemical and electrical sequences that shape who we are and what we feel, think, dream, and do. Although initially transient and instantaneous, enough repetition of any stimulus—whether it's operating a new technological device, or simply making a change in one's jogging route—will lay down a corresponding set of neural network pathways in the brain, which can become permanent.

Your brain—weighing about three pounds—sits cozily within your skull and is a complex mass of tissue, jam-packed with an estimated hundred billion cells. These billions of cells have central bodies that control them, which constitute the brain's *gray matter*, also known as the cortex, an extensive outer layer of cells or neurons. Each cell has extensions, or wires (axons) that make up the brain's *white matter* and connect to dendrites allowing the cells to communicate and receive messages from one another across synapses, or connection sites (Figure, page 6).

The brain's gray matter and white matter are responsible for memory, thinking, reasoning, sensation, and muscle movement. Scientists have mapped the various regions of the brain that correspond to different functions and specialized neural circuitry (Figure, page 7). These regions and circuits manage everything we do and experience, including falling in love, flossing our teeth, reading a novel, recalling fond memories, and snacking on a bag of nuts.

The amount and organizational complexity of these neurons, their wires, and their connections are vast and elaborate. In the average brain, the number of synaptic connection sites has been estimated at 1,000,000,000,000,000, or a million times a billion. After all, it's taken millions of years for the brain to evolve to this point. The fact that it has taken so long for the human brain to evolve such complexity makes the current single-generation, high-tech brain evolution so phenomenal. We're talking about significant brain changes happening over mere decades rather than over millennia.

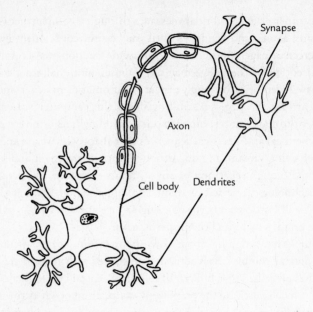

YOUNG PLASTIC BRAINS

The process of laying down neural networks in our brains begins in infancy and continues throughout our lives. These networks or pathways provide our brains an organizational framework for incoming data. A young mind is like a new computer with some basic programs built in and plenty of room left on its hard drive for additional information. As more and more data enter the computer's memory, it develops shortcuts to access that information. Email, word processing, and search engine programs learn the user's preferences and repeated keywords, for which they develop shortcuts, or macros, to complete words and phrases after only one or two keys have been typed. As young malleable brains develop shortcuts to access information, these shortcuts represent new neural pathways being laid down. Young children who have learned their times tables by heart no longer use the more cumbersome neural pathway of figuring out the math problem by counting their fingers or multiplying on paper. Eventually they learn even more effective shortcuts, such as ten times any number simply requires adding a zero, and so on.

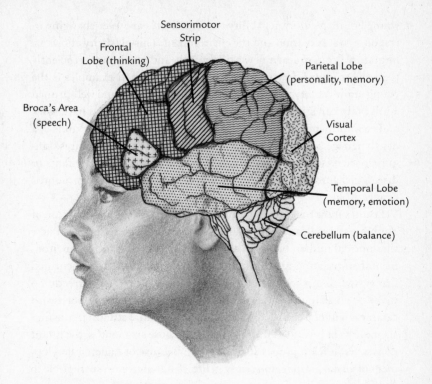

Frontal
Lobe (thinking)

Sensorimotor
Strip

Parietal Lobe
(personality, memory)

Broca's Area
(speech)

Visual
Cortex

Temporal Lobe
(memory, emotion)

Cerebellum (balance)

In order for us to think, feel, and move, our neurons or brain cells need to communicate with one another. As they mature, neurons sprout abundant branches, or dendrites, that receive signals from the long wires or axons of neighboring brain cells. The amount of cell connections, or synapses, in the human brain reaches its peak early in life. At age two, synapse concentration maxes out in the frontal cortex, when the weight of the toddler's brain is nearly that of an adult's. By adolescence, these synapses trim themselves down by about 60 percent and then level off for adulthood. Because there are so many potential neural connections, our brains have evolved to protect themselves from "over-wiring" by developing a selectivity and letting in only a small subset of information. Our brains cannot function efficiently with *too* much information.

The vast number of potentially viable connections accounts for the young brain's *plasticity*, its ability to be malleable and ever-changing in response to stimulation and the environment. This plasticity allows an immature brain to learn new skills readily and much more efficiently than the trimmed-down adult brain. One of the best examples is the young brain's ability to learn language. The fine-tuned and well-pruned adult brain can still take on a new language, but it requires hard work and commitment. Young children are more receptive to the sounds of a new language and much quicker to learn the words and phrases. Linguistic scientists have found that the keen ability of normal infants to distinguish foreign language sounds begins declining by twelve months of age.

Studies show that our environment molds the shape and function of our brains as well, and, it can do so to the point of no return. We know that normal human brain development requires a balance of environmental stimulation and human contact. Deprived of these, neuronal firing and brain cellular connections do not form correctly. A well-known example is visual sensory deprivation. A baby born with cataracts will not be able to see well-defined spatial stimuli in the first six months of life. If left untreated during those six months, the infant may never develop proper spatial vision. Because of ongoing development of visual brain regions early in life, children remain susceptible to the adverse effects of visual deprivation until they are about seven or eight years old. Although exposure to new technology may appear to have a much more subtle impact, its structural and functional effects are profound, particularly on a young, extremely plastic brain.

Of course, genetics plays a part in our brain development as well, and we often inherit cognitive talents and traits from our parents. There are families in which musical, mathematical, or artistic talents appear in several family members from multiple generations. Even subtle personality traits appear to have genetic determinants. Identical twins who were separated at birth and then reunited as adults have discovered that they hold similar jobs, have given their children the same names, and share many of the same tastes and hobbies, such as collecting rare coins or painting their houses green.

But the human genome—the full collection of genes that produces a human being—cannot run the whole show. The relatively modest number of human genes—estimated at twenty thousand—is tiny compared with the billions of synapses that eventually develop in our brains. Thus, the amount of information in an individual's genetic code would be insufficient to map out the billions of complex neural connections in the brain without additional environmental input. As a result, the stimulation we expose our minds to every day is critical in determining how our brains work.

NATURAL SELECTION

Evolution essentially means change from a primitive to a more specialized or advanced state. When your teenage daughter learns to upload her new iPod while IM'ing on her laptop, talking on her cell phone, and reviewing her science notes, her brain adapts to a more advanced state by cranking out neurotransmitters, sprouting dendrites, and shaping new synapses. This kind of moment-to-moment, day-in and day-out brain morphing in response to her environment will eventually have an impact on future generations through evolutionary change.

One of the most influential thinkers of the nineteenth century, Charles Darwin, helped explain how our brains and bodies evolve through *natural selection,* an intricate interaction between our genes and our environment, which Darwin simply defined as a "preservation of favorable variations and the rejection of injurious variations." Genes, made up of DNA—the blueprint of all living things—define who we are: whether we'll have blue eyes, brown hair, flexible joints, or perfect pitch. Genes are passed from one generation to the next, but occasionally the DNA of an offspring contains errors or mutations. These errors can lead to differing physical and mental attributes that could give certain offspring an advantage in some environments. For example, the genetic mutation leading to slightly improved visual acuity gave the "fittest" ancestral hunters a necessary advantage to avoid oncoming predators and go on to kill their prey. Darwin's principal of *survival of the fittest* helps explain how those with a genetic edge are more likely to survive, thrive, and pass their DNA on to the next generation. These DNA mutations also help explain the tremendous diversity within our species that has developed over time.

Not all brain evolution is about survival. Most of us in developed nations have the survival basics down—a place to live, a grocery store nearby, and the ability to dial 911 in an emergency. Thus, our brains are free to advance in creative and academic ways, achieve higher goals, and, it is hoped, increase our enjoyment of life.

Sometimes an accident of nature can have a profound effect on the trajectory of our species, putting us on a fast-track evolutionary course. According to anthropologist Stanley Ambrose of the University of Illinois, approximately three hundred thousand years ago, a Neanderthal man realized he could pick up a bone with his hand and use it as a primitive hammer. Our primitive ancestors soon learned that this tool was more effective when the other object was steadied with the opposite hand. This led our ancestors to develop right-handedness or left-handedness. As one side of the brain evolved to become stronger at controlling manual dexterity the opposite side became more specialized in the evolution of language. The area of the modern brain that controls the oral and facial muscle movement necessary for language—Broca's area—is in the frontal lobe just next to the fine muscle area that controls hand movement.

Nine out of ten people are right-handed, and their Broca's area, located in the left hemisphere of their brain, controls the right side of their body. Left-handers generally have their Broca's area in the right hemisphere of their brain. Some of us are ambidextrous, but our handedness preference for the right or the left tends to emerge when we write or use any hand-held tool that requires a precision grip.

In addition to handedness, the coevolution of language and tool making led to other brain alterations. To create more advanced tools, prehuman Neanderthals had to have a goal in mind and the planning skills to reach that goal. For example, ensuring that a primitive spear or knife could be gripped well and kill prey involved planning a sequence of actions, such as cutting and shaping the tool and collecting its binding material. Similar complex planning was also necessary for the development of grammatical language, including stringing together words and phrases and coordinating the fine motor lingual and facial muscles, which are thought to have further accelerated frontal lobe development.

In fact, when neuroscientists perform functional magnetic resonance imaging (MRI) studies while volunteers imagine a goal and carry out secondary tasks to achieve that goal, the scientists can pinpoint areas of activation in the most anterior, or forward, part of the frontal lobe. This frontal lobe region probably developed at the same time that language and tools evolved, advancing our human ancestors' ability to hold in mind a main goal while exploring secondary ones—the fundamental components of our human ability to plan and reason.

Brain evolution and advancement of language continue today in the digital age. In addition to the shorthand that has emerged through email and instant messaging, a whole new lexicon has developed through text messaging (see Chapter 8 and Appendix 2), based on limiting the number of words and letters used when communicating on hand-held devices. Punctuation marks and letters are combined in creative ways to indicate emotions, such as LOL = laugh out loud, and :-) = happy or good feelings. Whether our communications involve talking, written words, or even just emoticons, different brain regions control and react to the various types of communications. Language—either spoken or written—is processed in Broca's area in our frontal lobes. However, neuroscientists at Tokyo Denki University in Japan found that when volunteers viewed emoticons during functional MRI scanning, the emoticons activated the right inferior frontal gyrus, a region that controls nonverbal communication skills.

HONEY, DOES MY BRAIN LOOK FAT?

Natural selection has literally enlarged our brains. The human brain has grown in intricacy and size over the past few hundred thousand years to accommodate the complexity of our behaviors. Whether we're painting, talking, hammering a nail, or answering email, these activities require elaborate planning skills, which are controlled in the front part of the brain.

As Early Man's language and tool-making skills gradually advanced, brain size and specialization accelerated. Our ancestors who learned to use language began to work together in hunting groups, which helped them survive drought and famine. Sex-specific social roles evolved further as well. Males specialized in hunting, and those males with better visual and spatial abilities (favoring the right brain) had the hunting advantage. Our female ancestors took on the role of caring for offspring, and those with more developed language skills (left brain) were probably more nurturing to their offspring, so those offspring were more likely to survive. Even now, women tend to be more social and talk more about their feelings, while men, no longer hunters, retain their highly evolved right brain visual-spatial skills, thus often refusing to use the GPS navigation systems in their cars to get directions.

The printing press, electricity, telephone, automobile, and air travel were all major technological innovations that greatly affected our lifestyles and our brains in the twentieth century. Medical discoveries have brought us advances that would have been considered science fiction just decades ago. However, today's technological and digital progress is likely causing our brains to evolve at an unprecedented pace.

HIGH-TECH REVOLUTION AND THE DIGITAL AGE

Textile manufacturing, machine tools, steam power, railroads, and other technological discoveries were the driving forces behind the Industrial Revolution in the eighteenth and nineteenth centuries. Although not truly a revolution, since its gradual transformation spanned several hundred years, it changed the face of nations, gave rise to urban centers, created a middle class, and provided the economic base for a higher standard of living.

In 1961, two American electrical engineers, Jack Kilby and Robert Noyce, discovered something that led to our high-tech revolution—the silicon chip. This chip moved technology beyond the big and bulky vacuum tube, and even beyond the transistor, which required wired circuit boards. These engineers were able to combine components in an integrated circuit using silicon, a semiconductor material. This single innovation continues to rapidly advance our technology.

We've also witnessed the emergence of a new digital system of communication. The term *digital* essentially means any signal that is transmitted in a code of pluses and minuses, also known as a binary system. iPods and TiVos record and play back digitally. By contrast, record albums and tape recorders use an analog system, wherein the information is contained on a continuous surface that must be large enough to hold the length of the recording.

DO YOU REMEMBER . . .

- the first time you watched color TV?
- the 1961 introduction of the IBM Selectric typewriter, with its high-tech erase button?
- your first push button phone in the 1960s?
- your first remote control television set?
- Pong, the first video game?
- Sony's now obsolete Betamax video format of the late 1970s?
- the early mobile phones that required a suitcase to carry around?
- when you first started buying CDs instead of vinyl records or cassettes?

Our brain's neural circuits—axons, dendrites, and the synapses that connect them—are biologically primed to function digitally. For each thought or sensation—say, an itch on your right foot—multiple neurotransmitters are released from a neuron, and they all attempt to cross the synapse to communicate their information to the next neuron so the itch can get scratched. However, only a limited number of these neurotransmitters get through to the next neuron's receptor. Those that fail to connect signal a "0," while those that succeed in transmitting signal a "1." All the left-over zeros floating around represent the inefficiency of our brain's digital binary system. Essentially, neural processing is inefficient—the adult human brain accounts for 20 percent of our total energy expenditure. In other words, if you're eating a diet of two thousand calories per day, your brain alone burns up four hundred of those calories. Young developing brains require even more energy—a child's brain can use more than 50 percent of the entire body's caloric intake.

Despite the inefficiency of our basic biology, the brain, whether it's developing or fully matured, is able to adapt to newer and faster devices that are perpetually outdating the ones we already have. It seems as if your new computer or smart phone is already outdated before you can take it out of the box, and a newer, faster, more sophisticated model is sweeping the country.

To give this some perspective, think of how a single technological innovation—motion pictures—affected people's minds and expanded their sense of the world. Before newsreels and movies, most people were unable to directly observe or experience events outside of their own town and day-to-day lives. The advent of motion pictures and newsreels allowed people to witness a limitless range of experiences, whether it was bombs falling on the battlefields of Europe or the physical comedy of the Marx Brothers being chased through the corridors of a cruise ship. Movies had a profound social, political, and emotional impact on society. However, the effect on our brain wiring was relatively minimal because the exposure was limited. Most people went to the movies for only a couple of hours a week at most.

Now we are exposing our brains to technology for extensive periods each day, even at very young ages. A 2007 University of Texas study of more than a thousand children found that on a typical day, 75 percent of children watch TV, while 32 percent of them watch videos or DVDs, with a total daily exposure averaging one hour and twenty minutes. Of children who are five- and six-year-olds, an additional fifty minutes is spent in front of the computer.

A recent Kaiser Foundation study found that young people eight to eighteen years of age expose their brains to eight and a half hours of digital and video sensory stimulation each day. The investigators reported that most of the technology exposure is passive, such as watching television and videos (four hours daily) or listening to music (one hour and forty-five minutes), while other exposure is more active and requires mental participation, such as playing video games (fifty minutes daily) or using the computer (one hour).

YOUR BRAIN ON GOOGLE

We know that the brain's neural circuitry responds every moment to whatever sensory input it gets, and that the many hours people spend in front of the computer—doing various activities including trolling the Internet, exchanging email, video conferencing, IM'ing, and e-shopping—expose their brains to constant digital stimulation. Our UCLA research team wanted to look at how much impact this extended computer time was having on the brain's neural circuitry, how quickly it could build up new pathways, and whether or not we could observe and measure these changes as they occurred.

I enlisted the help of Drs. Susan Bookheimer and Teena Moody,

UCLA experts in neuropsychology and neuroimaging. We hypothesized that computer searches and other online activities cause measurable and rapid alterations to brain neural circuitry, particularly in people without previous computer experience.

To test our hypotheses, we planned to use functional MRI scanning to measure the brain's neural pathways during a common Internet computer task: searching Google for accurate information. We first needed to find people who were relatively inexperienced and naïve to the computer. Because the Pew Internet project surveys had reported that about 90 percent of young adults are frequent Internet users compared with less than 50 percent of older people, we knew that people naïve to the computer did exist and that they tended to be older.

After initial difficulty finding people who had not yet used computers, we were able to recruit three volunteers in their mid-fifties and sixties who were new to computer technology, yet willing to give it a try. To compare the brain activity of these three computer-naïve volunteers, we also recruited three computer-savvy volunteers of comparable age, gender, and socioeconomic background. For our experimental activity, we chose searching on Google for specific and accurate information on a variety of topics, ranging from the health benefits of eating chocolate to planning a trip to the Galapagos.

Next, we had to figure out a way to do MRI scanning on the volunteers while they used the Internet. Because the study subjects had to be inside a long narrow tube of an MRI scanner during the experiment, there would be no space for a computer, keyboard, or mouse. To re-create the Google-search experience inside the scanner, the volunteers wore a pair of special goggles that presented images of website pages designed to simulate the conditions of a typical Internet search session. The system allowed the volunteers to navigate the simulated computer screen and make choices to advance their search by simply pressing one finger on a small keypad, conveniently placed.

To make sure that the functional MRI scanner was measuring the neural circuitry that controls Internet searches, we needed to factor out other sources of brain stimulation. To do this, we added a control task that involved the study subjects reading pages of a book projected through the specialized goggles during the MRI. This task allowed us to subtract from the MRI measurements any nonspecific brain activations from simply reading text, focusing on a visual image, or concentrating. We wanted to observe and measure only the brain's activity from those mental tasks required for Internet searching, such as scanning for targeted key words, rapidly choosing from among several alternatives, going back to a previous page if a particular search choice was not helpful, and so forth. We alternated this control task—simply reading a simulated page of text—with the Internet searching task. We also controlled for nonspecific brain stimulations caused by the photos and drawings that are typically displayed on an Internet page.

Finally, to determine whether we could train the brains of Internet-naïve volunteers, after the first scanning session we asked each volunteer to search the Internet for an hour each day for five days. We gave the computer-savvy volunteers the same assignment and repeated the functional MRI scans on both groups after the five days of search-engine training.

As we had predicted, the brains of computer-savvy and computer-naïve subjects did not show any difference when they were reading the simulated book text; both groups had years of experience in this mental task, and their brains were quite familiar with reading books. By contrast, the two groups showed distinctly different patterns of neural activation when searching on Google. During the baseline scanning session, the computer-savvy subjects used a specific network in the left front part of the brain, known as the dorsolateral prefrontal cortex. The Internet-naïve subjects showed minimal, if any, activation in this region.

One of our concerns in designing the study was that five days would not be enough time to observe any changes, but previous research suggested that even Digital Immigrants can train their brains relatively quickly. Our initial hypothesis turned out to be correct. After just five days of practice, the exact same neural circuitry in the front part of the brain became active in the Internet-naïve subjects. Five hours on the Internet, and the naïve subjects had already rewired their brains. The following figure shows the neural network (arrows) that a Google search will trigger after just a few days of activity on the computer.

This particular area of the brain controls our ability to make decisions and integrate complex information. It also controls our mental

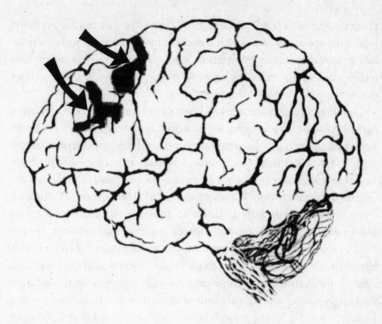

process of integrating sensations and thoughts, as well as working memory, which is our ability to keep information in mind for a very short time—just long enough to manage an Internet search task or dial a phone number after getting it from directory assistance.

The computer-savvy volunteers activated the same frontal brain region at baseline and had a similar level of activation during their second session, suggesting that for a typical computer-savvy individual, the neural circuit training occurs relatively early and then remains stable. But these initial findings raise several unanswered questions. If our brains are so sensitive to just an hour a day of computer exposure, what happens when we spend more time? What about the brains of young people, whose neural circuitry is even more malleable and plastic? What happens to their brains when they spend their average eight hours daily with their high-tech toys and devices?

TECHNO-BRAIN BURNOUT

In today's digital age, we keep our smart phones at our hip and their earpieces attached to our ears. A laptop is always within reach, and there's no need to fret if we can't find a land line—there's always Wi-Fi (short for *wireless fidelity*, which signifies any place that supplies a wireless connection to the Internet) to keep us connected. As technology enables us to cram more and more work into our days, it seems as if we create more and more work to do.

Our high-tech revolution has plunged us into a state of *continuous partial attention*, which software executive Linda Stone describes as continually staying busy—keeping tabs on everything while never truly focusing on anything. Continuous partial attention differs from multitasking, wherein we have a purpose for each task, and we are trying to improve efficiency and productivity (see Chapter 7). Instead, when our minds partially attend, and do so continuously, we scan for an opportunity for any type of contact at every given moment. We virtually chat as our text messages flow, and we keep tabs on active buddy lists (friends and other screen names in an instant message program); everything, everywhere is connected through our peripheral attention. Although having all our pals online from moment to moment seems intimate, we risk losing personal touch with our real-life relationships and may experience an artificial sense of intimacy compared with when we shut down our devices and devote our attention to one individual at a time. But still, many people report that if they're suddenly cut off from someone's buddy list, they take it personally—deeply personally.

When paying partial continuous attention, people may place their brains in a heightened state of stress. They no longer have time to reflect, contemplate, or make thoughtful decisions. Instead, they exist in a sense of constant crisis—on alert for a new contact or bit of exciting news or information at any moment. Once people get used to this state, they tend to thrive on the perpetual connectivity. It feeds their egos and sense of self-worth, and it becomes irresistible.

Neuroimaging studies suggest that this sense of self-worth may protect the size of the hippocampus—that horseshoe-shaped brain region in the medial (inward-facing) temporal lobe, which allows us to learn and remember new information. Dr. Sonia Lupien and associates at McGill University studied hippocampal size in healthy younger and older adult volunteers. Measures of self-esteem correlated significantly with hippocampal size, regardless of age. They also found that the more people felt in control of their lives, the larger the hippocampus.

But at some point, the sense of control and self-worth we feel when we maintain partial continuous attention tends to break down—our brains were not built to maintain such monitoring for extended time periods. Eventually, the endless hours of unrelenting digital connectivity can create a unique type of brain strain. Many people who have been working on the Internet for several hours without a break report making frequent errors in their work. Upon signing off, they notice feeling spaced out, fatigued, irritable, and distracted, as if they are in a "digital fog." This new form of mental stress, what I term techno-brain burnout, is threatening to become an epidemic.

Under this kind of stress, our brains instinctively signal the adrenal gland to secrete cortisol and adrenaline. In the short run, these stress hormones boost energy levels and augment memory, but over time they actually impair cognition, lead to depression, and alter the neural circuitry in the hippocampus, amygdala, and prefrontal cortex—the brain regions that control mood and thought. Chronic and prolonged techno-brain burnout can even reshape the underlying brain structure.

Dr. Sara Mednick and colleagues at Harvard University were able to experimentally induce a mild form of techno-brain burnout in research volunteers; they then were able to reduce its impact through power naps and by varying mental assignments. Their study subjects performed a visual task: reporting the direction of three lines in the lower left corner of a computer screen. The volunteers' scores worsened over time, but their performance improved if the scientists alternated the visual task between the lower left and lower right corners of the computer screen. This result suggests that brain burnout may be relieved by varying the location of the mental task.

The investigators also found that the performance of study subjects improved if they took a quick twenty- to thirty-minute nap. The neural networks involved in the task were apparently refreshed during rest; however, optimum refreshment and reinvigoration for the task occurred when naps lasted up to sixty minutes—the amount of time it takes for rapid eye movement (REM) sleep to kick in.

THE NEW, IMPROVED BRAIN

Young adults have created computer-based social networks through sites like MySpace and Facebook, chat rooms, instant messaging, video conferencing, and email. Children and teenagers are cyber-savvy too. A fourteen-year-old girl can chat with ten of her friends at one time with the stroke of a computer key and find out all the news about who broke up with whom in seconds—no need for ten phone calls or, heaven forbid, actually waiting to talk in person the next day at school.

These Digital Natives have defined a new culture of communication—no longer dictated by time, place, or even how one looks at the moment unless they're video chatting or posting photographs of themselves on MySpace. Even baby boomers who still prefer communicating the traditional way—in person—have become adept at email and instant messaging. Both generations—one eager, one often reluctant—are rapidly developing these technological skills and the corresponding neural networks that control them, even if it's only to survive in the ever-changing professional world.

Almost all Digital Immigrants will eventually become more technologically savvy, which will bridge the brain gap to some extent. And, as the next few decades pass, the workforce will be made up of mostly Digital Natives; thus, the brain gap as we now know it will cease to exist. Of course, people will always live in a world in which they will meet friends, date, have families, go on job interviews, and interact in the traditional face-to-face way. However, those who are most fit in these social skills will have an adaptive advantage. For now, scientific evidence suggests that the consequences of early and prolonged technological exposure of a young brain may in some cases never be reversed, but early brain alterations can be managed, social skills learned and honed, and the brain gap bridged.

Whether we're Digital Natives or Immigrants, altering our neural networks and synaptic connections through activities such as email, video games, Googling (verb: to use the Google search engine to obtain information on the Internet [from *Wikipedia; the free encyclopedia*]), or other technological experiences does sharpen some cognitive abilities. We can learn to react more quickly to visual stimuli and improve many forms of attention, particularly the ability to notice images in our peripheral vision. We develop a better ability to sift through large amounts of information rapidly and decide what's important and what isn't—our mental filters basically learn how to shift into overdrive. In this way, we are able to cope with the massive amounts of information appearing and disappearing on our mental screens from moment to moment.

Initially, the daily blitz of data that bombards us can create a form of attention deficit, but our brains are able to adapt in a way that promotes rapid information processing. According to Professor Pam Briggs of North Umbria University in the United Kingdom, Web surfers looking for information on health spend two seconds or less on any particular website before moving on to the next one. She found that when study subjects did stop and focus on a particular site, that site contained data relevant to the search, whereas those they skipped over contained almost nothing relevant to the search. This study indicates that our brains learn to swiftly focus attention, analyze information, and almost instantaneously decide on a go or no-go action. Rather than simply catching "digital ADD," many of us are developing neural circuitry that is customized for rapid and incisive spurts of directed concentration.

While the brains of today's Digital Natives are wiring up for rapid-fire cyber searches, the neural circuits that control the more traditional learning methods are neglected and gradually diminished. The pathways for human interaction and communication weaken as customary one-on-one people skills atrophy. Our UCLA research team and other scientists have shown that we can intentionally alter brain wiring and reinvigorate some of these dwindling neural pathways, even while the newly evolved technology circuits bring our brains to extraordinary levels of potential.

Although the digital evolution of our brains increases social isolation and diminishes the spontaneity of interpersonal relationships, it may well be increasing our intelligence in the way we currently measure and define IQ. Average IQ scores are steadily rising with the advancing digital culture, and the ability to multitask without errors is improving. Neuroscientist Paul Kearney at Unitec in New Zealand reported that some computer games can actually improve cognitive ability and multitasking skills. He found that volunteers who played the games eight hours each week improved multitasking skills by two and a half times. Other research at Rochester University has shown that video game playing can improve peripheral vision as well. As the modern brain continues to evolve, some attention skills improve, mental response times sharpen, and the performance of many brain tasks becomes more efficient. These new brain proficiencies will be even greater in future generations and alter our current understanding and definition of intelligence.

TAKING CONTROL OF YOUR BRAIN'S EVOLUTION

You can get a better sense of how your own brain is adapting to the high-tech revolution, and begin to take control of your neural circuitry by making informed choices about the quantity and quality of your brain's technological exposure. At the same time you'll discover how the digital age is affecting your offline traditional brain stimulation, and in what areas you need to train your brain in order to succeed in this changing environment.

All of us, Digital Natives and Immigrants, will master new technologies and take advantage of their efficiencies, but we also need to maintain our people skills and humanity. Whether in relation to a focused Google search or an empathic listening exercise, our synaptic responses can be measured, shaped, and optimized to our advantage, and we can survive the technological adaptation of the modern mind.

Two

BRAIN GAP:
Technology Dividing Generations

> *That which seems the height of absurdity in one generation often becomes the height of wisdom in another.*
>
> *Adlai Stevenson*

You look at the box your husband and teenage daughter gave you last Christmas, and even though it's almost Labor Day, that darn computer is still untouched inside it. After all, you've been an author for more than twenty-five years, and you happen to like writing in longhand. And who do you have to email, anyway? Your agent? Your editor? Your publicist? They can all just spend a dime and call you. So what if your daughter laughs at you and calls you technologically challenged and your husband accuses you of being scared to take the digital leap? You will *not* be pressured into anything.

It's a year later, and you write strictly on the computer now. Okay, so it *is* more efficient once you get the hang of it. And downloading your schedule from a hand-held to the desktop keeps everything nice and co-ordinated. So maybe your husband and daughter were right, but they'll never hear it from you. Hold on—what's happening? Why can't you type anything? Why won't the cursor move? The buttons are dead! The whole damn keyboard is frozen, and you have an entire manuscript on this idiotic computer! Not to mention your schedule, phone book, and emails to and from everyone you know—DAMMIT! You knew these machines were garbage!

Your daughter hears you yelling from the other room and runs in. You fling yourself on the sofa, proclaiming that your life is over. She sits at your desk for three seconds and then looks at you as if you're crazy, "Mom, have you ever heard of changing the batteries in your keyboard?"

Today's dizzying pace of high-tech innovation not only presents a challenge for those of us raised before there was a computer on every desk, but is actually altering the neural wiring of tech-savvy young people's brains—changing the way they develop and function, and turning the normal generation gap into something new: a widening chasm I call the brain gap. Our society appears to be breaking into two cultural groups: Digital Natives, who were born into a world of computer technology, and Digital Immigrants, who were introduced to computer technology as adults.

In the past, young people tended to rebel against their parents' morals for a while and then eventually integrate themselves into their parents' society—adopting much of the work ethics, attitudes, and values of the older generation while bringing their own culture, outlooks, and perspective into the mix. But today young digital minds are adapting to a new technology-driven culture that is overtaking yesterday's low-tech lifestyle. This younger generation is jettisoning their parents' values and establishing a new social and political network, instituting their own cyberspace manners, language, and workplace ethics into the mainstream.

Many baby boomers, in their forties and older, have experienced the generation gap, not just with their own parents but with their kids as well—perhaps when telling their teenagers some of the same things their parents said to them, such as "You call *that* music?" or "You're not leaving this house wearing *that*, young lady." (I once uttered the dreaded "Because I said so!") However, the brain gap refers to much more than intergenerational differences in tastes and values. It points to an actual evolutionary change in the wiring of today's younger minds—a change in neural circuitry that is fundamentally different from that of their parents and grandparents.

DIGITAL NATIVES

The younger generation of Digital Natives has grown up immersed in technology that continually becomes more powerful and compact— literally, cyberspace in their pockets. They multitask and parallel process with ease, and their access to visual and auditory stimulation has programmed their brains to crave instant gratification. Neuroscientists at Princeton University have found that our brains use different regions to balance short-term and long-term rewards. When we make decisions that instantly gratify our needs, the brain's emotional centers in the limbic system take over. But those regions have trouble thinking ahead to the future, and neural circuits in the brain's centers of logic in the frontal lobe and parietal cortex are required for us to put off a reward.

The bombardment of digital stimulation on developing minds has taught them to respond faster, but they encode information differently from the way older minds do. Digital Natives tend to have shorter attention spans, especially when faced with traditional forms of learning. This young high-tech generation often finds conventional television too sluggish and boring when simply watched on its own. One-third of young people use other media—particularly the Internet—while watching television. Even middle-school students multitask almost constantly, downloading music to their iPods and instant messaging their friends while doing their homework. Their young developing brains are much more sensitive to environmental input than are more mature brains. Ironically, it's the younger minds that not only are the most vulnerable to the brain-altering influence of new technology but also are the most exposed to it.

Young people today spend much less time reading for leisure than ever before (see Chapter 1). After all, why spend time staring at a dull and stagnant string of words when they could be entertained and informed with fast-paced visual and auditory computer images instead? Some Digital Natives also complain that books make them feel isolated—they want to stay connected with their friends online instead of holing up alone with a book in the bedroom or the library.

Technological advances have brought many new ways of learning into the classroom and the home. Online courses are available for high-schoolers, college students, and adults. Search engines such as Yahoo and Google provide vast resources for research on almost every subject. Young children begin using the computer in preschool or earlier, and numerous computer programs, such as Kurzweil, Leapfrog, Fast Forward, and Draft:Builder (see Appendix 3), are designed to help kids learn to read and write earlier and to develop their hand-eye coordination at a younger age. They also prepare kids to multitask more

effectively. However, recent studies suggest that too much video exposure, even to these so-called educational videos, can delay language
development in young children.

It's a Smaller, Smaller World

Because of new technology and globalization, young Digital Natives
are experiencing a shrinking world. With 24/7 access to almost anything and anyone, Internet, email, and instant messaging have become
the communication modes of choice for many people—young and old.
People blog, students keep in touch with teachers, colleagues communicate rapidly with colleagues, and friends drop quick notes to one another. Even the traditional party invitation is being replaced by the
e-vite.

   The workplace is becoming more efficient. Cyberspace meetings can
be held among multiple executives around the globe, search engines
have brought a world of data and statistics down to just a keystroke,
and social networks have been created to enrich and expand the amount
of communication, information sharing, and entertainment available
in microseconds. MySpace, YouTube, Internet dating, and Web-based
shopping have all made people's lives more convenient, entertaining,
and faster paced than ever before. Globalization and outsourcing of
business resources is occurring around the world, in real time. You may
call to make a reservation at a local New York City restaurant and actually be speaking to someone in India who asks if you'd like a table with
a view of the park.

   While the brains of today's young Digital Natives are wiring up for
rapid-fire cyber searches, the neural circuitry and some parts of the
brain that normally adapt to more traditional learning methods are
becoming less developed. Many students acknowledge that classroom
learning and the customary lecture/note-taking system seem boring to
them. Most teens no longer write in personal diaries but rather share
their innermost thoughts with friends—and often strangers—on websites and blogs. They think nothing of tossing out a digital device they
recently bought in order to upgrade to a newer one with a clearer image,
faster speed, better keyboard, or higher-capacity memory—especially if
it looks cool.

Sensitive Developing Minds

Even before birth, a baby's brain health is extremely susceptible to its mother's lifestyle habits. Drinking alcohol could put a baby at risk for fetal alcohol syndrome, the most common preventable cause of mental retardation. Smoking cigarettes during pregnancy may inhibit prenatal brain growth. Mothers who don't ingest enough folic acid, especially before pregnancy occurs, can have babies with neural tube defects, and even emotional stress during pregnancy can impair a newborn's coordination, response times, and ability to focus attention.

Most of a baby's brain synapses are formed during the first six months of life, when the brain consumes more than 60 percent of the body's total caloric intake. Too little brain stimulation during this period will lead to the formation of fewer synapses; too much could lay down faulty synapses and maladaptive neural circuits.

Toddlers and small children model the behavior of their parents, other adults, and peers as they listen and learn to pay attention, as well as communicate and interact socially. Reading to a child daily, expressing affection frequently, and other nurturing interactions stimulate the young child's brain so that new dendrites grow and branch out toward one another. Functional MRI and positron emission tomography (PET) scans show that specific neural circuits are preprogrammed to be activated when young children pay attention to other children and adults.

Without enough face-to-face interpersonal stimulation, a child's neural circuits can atrophy, and the brain may not develop normal interactive social skills. However, overstimulation can affect a child's brain development negatively as well. Too many extracurricular activities, too much tutoring, or a home environment that is extremely chaotic can overwhelm a child's developing neural circuitry, leading to low self-esteem, anxiety, and distractibility. When a child's brain is exposed to excessive levels of television, computer, video, and other digital stimulation, it can lead to hyperactivity, irritability, and attention deficit disorders (see Chapter 4). The American Academy of Pediatrics actually recommends *no* television or video watching for children under two years of age.

As infants mature, their brains become less sensitive to outside stimulation, but children and adolescents still have many developmental milestones to achieve on their way to adulthood. The nineteenth-century French psychologist Jean Piaget charted these milestones to adulthood (see Table), beginning with the first two years of life, when a toddler develops awareness of other people and learns to relate to them. From two to six years, the young child learns basic language skills. However, thinking is relatively concrete until the teen years, when the ability for abstract thought and reason takes hold. If digital technology continues to distract young susceptible minds at the present rate, the traditional developmental stages will need to be redefined.

JEAN PIAGET'S DEVELOPMENTAL STAGES

STAGE/MILESTONE	AGE
Sensorimotor Experience world through looking, touching, mouthing, early language	Birth to 2 years
Preoperational Use words and images to represent things	to 6 years
Concrete operational Think logically about concrete events	to 12 years
Formal operational Reason abstractly	to 19 years

Young Brains on Overdrive

As teenagers move into their twenties, a high percentage of them continue to overexpose their still malleable brains to complicated digital technology. A study commissioned by Microsoft Corporation found that younger age groups are much more likely to use computers than are older groups: more than 80 percent of people in their twenties use computers, compared with less than 30 percent of those over age seventy-five (see Chart), but older adults have been striving to catch up. The study projected that within a decade, twice as many people in their late sixties and early seventies will use computers, compared with today.

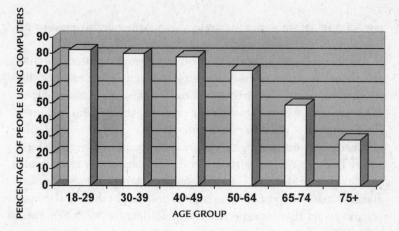

Youth dominates not only in computer use but also in using the Internet and other digital communication methods. A recent Pew Internet survey found that three out of four Americans frequent the Web, but teens and young adults constitute the highest proportion of users: approximately 90 percent of young adults go online, versus a mere one-third of people sixty-five years old or older.

It's not just the frequency of technology use but the *type* of use that separates the generations. Young people are more likely to use instant messaging: more than 60 percent of young adults use instant messaging, compared with only about 30 percent of older people. Approximately 40 percent of people in their thirties or younger send photos, humorous articles, or website links to others, whereas only about 20 percent of older adults exchange such electronic information.

Other research confirms that young people are spending more and more time exposing their brains to *all* forms of new media. A 2005 Kaiser Foundation and Stanford University study of more than two thousand kids and teens aged eight to eighteen years found that total daily media exposure had increased over the previous five years from seven hours twenty-nine minutes to eight hours thirty-three minutes. Today's adolescents are now spending more than a full eight-hour work day exposing their brains to digital technology. By spending this much time staring at a computer or television screen, these young people are not solidifying the normal neural pathways their brains need to develop traditional face-to-face communication skills.

An estimated 20 percent of this younger generation meets the clinical criteria for pathological Internet use—they are online so much that it interferes negatively with almost every other aspect of their lives. Their excessive Web use lowers their academic achievement and interferes with their social lives (see Chapter 3).

Whether or not their digital time reaches compulsive levels, the sedentary hours in front of a computer or television screen affect young people's physical health. In a 2006 study, Naoko Koezuka and associates at the University of Toronto studied nearly eight thousand teenagers and found that, as expected, the more time the volunteers spent playing video games, using the computer, and watching television, the less likely they were to spend time engaging in physical exercise. A recent study of children five to eleven years old found that those who watched more than one hour of television each day had increased body weight compared with children who watched less.

Empathy and the Adolescent Brain

Adolescence is a critical stage of development—a time when the brain advances from concrete to abstract thinking. This is traditionally when teenagers develop their capacity to understand the emotional experience of others, as well as learn and practice their empathic skills. Spending hours staring at a computer or video screen and perhaps listening to blaring music through their headphones at the same time likely hinders the development of adequate brain circuitry needed to accomplish these milestones.

Dr. Robert McGivern and co-workers at San Diego State University have found that when kids enter adolescence, they struggle with the ability to recognize another person's emotions. During the study, the teenage volunteers viewed faces demonstrating different emotional states. Compared with other age groups, eleven- and twelve-year-olds (the age when puberty typically kicks in) needed to take more time to identify the specific emotions expressed by the faces presented to them. It took longer for their frontal lobes to identify happy, angry, or sad faces, because of the pruning or trimming down of excess synaptic connections that occurs during puberty. However, once that pruning-down process is complete and the teenager matures to adulthood, expression recognition becomes faster and more efficient.

Scientists have pinpointed a specific region of the teenage brain that controls this tendency toward selfishness and a lack of empathy. Dr. Sarah-Jayne Blakemore of University College in London used functional MRI scanning to study the brains of teenagers (eleven to seventeen years) and young adults (twenty-one to thirty-seven years) while they were asked to make everyday decisions, such as when and where to see a movie or go out to eat. The scientists found that teenagers, when making these choices, used a brain network in their temporal lobes (underneath the temples), while older volunteers used the prefrontal cortex—a region that processes how our decisions affect other people. Such differing neural circuitry may explain why teens are less able to appreciate how their decisions affect those around them.

The same scientists further assessed how rapidly teenagers could consider the impact of their decisions on another person's welfare. The teens in the study were asked questions such as this: "How would your friend feel if she weren't invited to your party?" Younger volunteers took much longer to answer such questions. As people get older they are more able to put themselves in another person's shoes by using the neural circuitry in their frontal lobes.

Blakemore hypothesized that an evolutionary explanation may account for this age effect on decision making. In early times, adult humans with empathic skills had an adaptive advantage by forming groups that could fend off predators together and hunt for prey more successfully. Their young offspring had less need for empathy, since they were still being cared for by adults. Young people probably began considering other people's perspectives when they got older, and such decision making affected their survival.

Teenagers desire instant gratification—they want to satisfy their needs and do it now, not later. Their underdeveloped frontal lobes often impair their everyday judgment. Many teens feel they are invincible—danger will bounce off them. With normal maturation and age, the frontal lobe neural circuits strengthen, and judgment improves. We develop a greater capacity to delay gratification, consider other people's feelings, put things into perspective, and understand the danger certain situations may hold.

Unfortunately, today's obsession with computer technology and video gaming appears to be stunting frontal lobe development in many teenagers, impairing their social and reasoning abilities. If young people continue to mature in this fashion, their brains' neural pathways may never catch up. It is possible that they could remain locked into a neural circuitry that stays at an immature and self-absorbed emotional level, right through adulthood.

R U IM'ing While U Read This?

Multitasking originally referred to a computer's ability to carry out several tasks simultaneously. We now use it to describe a condition wherein people juggle multiple tasks at once, as opposed to completing one task before moving on to another task in a linear fashion (see also Chapter 4). Digital Natives love to multitask, and they're good at it. If a teenager were reading this book, he might well be instant messaging his friends and listening to his iPod at the same time. But studies show that too much multitasking can lead not only to increased stress and attention deficits but also to a decline in work efficiency.

Digital Natives are much more likely to multitask than are Digital Immigrants. In 2006, a *Los Angeles Times*/Bloomberg poll gathered responses from 1,650 volunteers and found that the majority of teenagers were busy with other things while they were doing their homework: 84 percent of them listened to music while studying, 47 percent watched TV, and 21 percent were doing three or more tasks at once.

Multitasking allows Digital Natives to instantly gratify themselves and put off long-term goals. The competing simultaneous tasks often provide a superficial view of the information being presented rather than in-depth understanding. Educators complain that young people in this multitasking generation are less efficient in their school work. Chronic and intense multitasking may also delay adequate development of the frontal cortex, the area of the brain that helps us see the big picture, delay gratification, reason abstractly, and plan ahead. If a teenager has the tools and know-how to gain immediate mental gratification from instant messaging or playing a video game, when will that teen learn to delay satisfying every pressing whim or urge in order to completely finish a tedious project or dull task?

For most Digital Immigrants, extreme multitasking is an inefficient way to use the brain—their original neural circuitry was not set up for it. Professor Patricia Tun of Brandeis University in Massachusetts found that simultaneous competing tasks can negatively affect a baby boomer's everyday life. Her studies showed, among other things, that as we age we have much more difficulty understanding rapid speech when we are distracted by background noise.

Typically, younger Digital Natives can handle three or more tasks simultaneously (e.g., email, telephone, iPod download), whereas a Digital

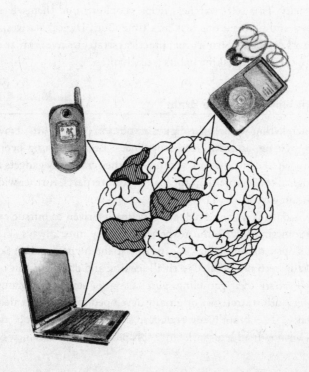

Immigrant's brain functions better when focused on one task at a time. The truth is that everyone is somewhat challenged by multitasking. A recent UCLA experiment showed that when volunteers between eighteen and forty-five years old were given a learning task while being asked to keep a count of distracting beep sounds, their recall scores dropped dramatically in comparison with their performance when they were not distracted.

In today's world, people tend to multitask at every age. Some middle-aged executives have placed a computer above their treadmills so they can get their cardiovascular workout while they catch up with their business email. Many states have recognized some of the obvious hazards of multitasking and are following New York's lead in banning drivers from talking on their hand-held cell phones. At what point does multitasking reduce our efficiency, create stress, and place us in danger? It depends on the individual—on how much the brain has been trained to multitask, as well as the individual's awareness of his or her own capacity. This self-awareness helps people remind themselves to slow down and perform one task at a time. Both Digital Natives and Immigrants can benefit from other practical strategies to optimize and minimize their multitasking habits (see Chapter 7).

Wanted: Novelty for the Brain

One principle that propels the digital revolution is our brain's craving for new, exciting, and different experiences. Because young people's brains are wired differently for technology, they crave new gadgets and technological tools more than their older counterparts, further widening the brain gap.

Some people, young and old, are genetically driven to pursue exciting and sometimes dangerous experiences. They must always ski the steepest slopes, drive the fastest cars, or bet the highest stakes. Some can actually become addicted to thrill seeking and often use new technology to satisfy their gambling, purchasing, or sex drives. Many require intervention strategies originally developed for drug and alcohol dependence. The brain neurotransmitter dopamine controls these drives. This powerful brain messenger propels people to seek new environments and experiences. The dopaminergic urge can be so strong that individuals often ignore discomfort or pain while attempting to satisfy their desire for novelty (see Chapter 3).

Whether excessive or subtle, the instinct to pursue new and exciting experiences frequently drives our behavior. The latest technology can be so enticing, so stimulating, and so much fun that before we know it, hours have passed. People of all ages are drawn in and fascinated by high-tech advances. Each new gadget they see in a catalogue or magazine, whether it's goggles displaying movie downloads or a voice-activated word-processing program, piques their interest—but not everyone *has* to have *that* one, right *NOW*.

The high-tech revolution has accelerated our compulsion to pursue what's newer and better to the point where much of our youth-oriented culture obsessively avoids technology obsolescence. Our environment's incessant digital bombardment has caused young brains to evolve in such a way that each technological invention has an almost irresistible draw. By contrast, Digital Immigrants often find that new technological advances pose yet another aggravating chore: they must again learn a new operating program, or buy yet another gadget that will probably become obsolete within months.

It's no surprise that these two generations differ in their electronic purchasing patterns. Digital Natives are compelled to buy the hottest and the latest devices—they will happily replace perfectly functioning equipment with a new innovation because it's faster, clearer, more powerful, bigger, or smaller—whatever is in fashion. When Digital Immigrants purchase a new device, they often plan to use it until it wears out. They "future proof" their computer purchases by choosing those with the most memory and greatest capacity to upgrade, rather than replace entire devices. Digital Immigrants tend to like the familiar—it's comfortable even if it is not optimal. Many hesitate before buying the newest equipment, knowing that they have little time or patience to even read the manual.

Video Game–Brain

The brain's craving for novelty has spurred a $10 billion video gaming industry. Video games have become so popular that they are becoming a spectator sport in some parts of the world. Cyber athletes compete before crowds of a hundred thousand or more in South Korea and elsewhere in video game tournaments.

In studies at Tokyo's Nihon University, Professor Akio Mori found evidence that video games appear to suppress frontal lobe activity. His group showed that the more time adolescents spend playing video games, the less time they use key areas of the front parts of their brains. Chronic players—those who play two to seven hours each day—sometimes develop video game–brain, a syndrome that essentially turns off the frontal lobes, even when the kids are not playing video games.

Gamers tend to get wrapped up in their play, forgetting or ignoring what else is going on around them. The player's increased physical and psychological arousal frequently leaves them feeling tense and irritable. Video gaming has been found to increase blood pressure and heart rate and to whip up the body's autonomic nervous system, cranking out stress-related chemical messengers, such as adrenaline, usually released by the adrenal glands in response to danger. Thus, chronic gaming may have negative consequences for the body as well as the brain.

Digital Natives constitute the major market for video gaming: more than 90 percent of all children and adolescents in the United States play these games. Although the stereotype of a fifteen-year-old cyber-geek gaming for hours in his bedroom persists, the average age of video gamers has risen to thirty years. Younger players, however, are the ones whose brains are most sensitive to extensive game play, and on average, these kids and teens are playing video games a half hour or more each day.

Previous research has shown that extensive video gaming makes kids more aggressive and desensitizes them to violence they see elsewhere. One study found that as little as ten minutes of daily violent video gaming could increase aggressive traits and behaviors. Some studies have downplayed the aggressive effects of video gaming. However, recent investigations suggest that the *intensity* of a game's violent graphics, rather than the *amount* of violent content, may have a greater effect on brain function and aggressive behavior.

Today, many video games put less emphasis on violence and destruction and focus more on complex goals and strategies. These intricate virtual environments have a significant impact on a young person's brain in the frontal lobe—the region that requires further development during adolescence for abstract thinking and planning skills to take hold. Professor Ryuta Kawashima and colleagues at Tohoku University in Japan found that when children play video games, their brains do not use frontal lobe circuits but rather use a limited brain region that controls vision and movement. This is in sharp contrast to what they found when kids performed simple, mundane math exercises. When the study volunteers did single-digit addition calculations, their brains recruited neurons from a much wider brain area, involving the frontal lobes, regions that control learning, memory, emotion, and even impulse control.

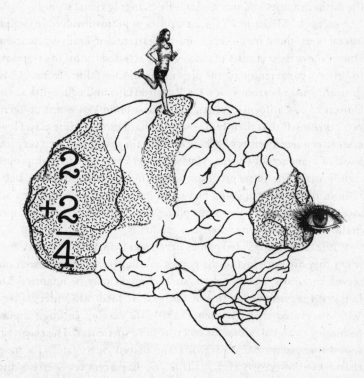

I Can Quit Any Time I Want

Video games are often intellectually and emotionally seductive, and they can easily become addictive, driving the brain's dopamine circuitry so that the player craves more and more play. This habit-forming element is present in many forms of digital technology, particularly those that have an interactive component, including email, instant messaging, Web shopping, MySpace, YouTube, eBay, Google or Yahoo searching, Internet gambling, and video gaming. Users or players have the opportunity to perpetually challenge themselves to search more or multitask faster, thus avoiding becoming bored or disinterested. Low-tech games and brain teasers—crossword puzzles, Sudoku, Scrabble—also provide a mental challenge that can be exhilarating, but the computer accelerates and intensifies these stimulating brain effects. Young Digital Natives have become stimulus junkies, drawn into the flashy graphics and intense, rapidly changing visual stimuli.

Functional MRI and PET scan studies of brains involved in digital interactions show remarkably similar patterns of brain adaptation. One of the earliest studies of video games assessed the brain's response to Tetris, a game requiring the player to maneuver falling rectangles to fit neatly into a bottom space. Dr. Richard Haier and colleagues at the University of California, Irvine, found that brain PET scans of Tetris novices showed that their brains worked hard during game play. However, after a few months of practice, the brain scans indicated very low levels of brain activity during play, while performance scores improved significantly. The Tetris program, like many other digital games, has a built-in difficulty gauge—as players improve their abilities, the pace picks up to further challenge the player. Brain efficiency increases optimally when the challenge increases gradually.

While teens using video games have lower frontal lobe activity even when they are not playing the games, similar patterns have been observed in adults, but for them such changes may be adaptive. Dr. James Rosser and associates of Beth Israel Medical Center in New York discovered that the mental skills we develop through digital technology can carry over to our real-life experiences. The scientists found that laparoscopic surgeons who played video games for more than three hours each week made nearly 40 percent fewer errors during surgical procedures, compared with surgeons who did not play video games.

The visual and mental stimulation of playing a video game is not unlike searching a Web browser or sorting through and responding to dozens of emails—two typical daily computer activities. The right balance of video gaming and digital interaction has the potential to quicken reaction time and improve some forms of attention, particularly the ability to notice peripheral visual stimuli. Our email and Web-searching sessions may even help us develop our ability to sift through large amounts of information rapidly.

But what about a passive video activity like watching television? Does that engage fewer brain neural circuits, leaving unused areas to shrivel away from neglect? The video game research findings emphasize the important influence of such digital technologies on brain development as well as behavior while players are not gaming. Are we rearing a new generation with underdeveloped frontal lobes—a group of young people unable to learn, remember, feel, or control their impulses? Or will the Digital Natives develop new advanced skills that will poise them for extraordinary achievements?

The answers are varied. Video games are not intrinsically bad—people of all ages have fun playing them. And other digital technologies provide extraordinary tools for communication, business, and social exchange. The issue is how much gaming, emailing, and Googling is too much for the mind—for a developing brain as well as a mature one. We each need to discover our optimum level of brain stimulation at every stage of life.

Although scientists have not yet determined the specific answers to these questions, we do know that a limited amount of video gaming may enrich minds and improve some forms of cognitive performance, while too much can lead people to become spaced-out video game-brainers who are unresponsive to the real world around them. Video gaming in moderation can help develop improved pattern recognition, more systematic thinking, and better executive skills. However, to truly succeed in games and in our lives, we need to learn to become more patient and delay gratification—both important adolescent developmental milestones. Some advocates argue that video games may very well exercise the mind just as physical conditioning exercises the body.

Scientists and game designers are developing new uses for video games that teach standard school curricula and other useful cognitive skills. Game and toy companies have begun marketing games for baby boomers to exercise their brains and improve memory performance (see Chapter 8).

DIGITAL IMMIGRANTS

While Digital Natives remain plugged into cyberspace and video games, Digital Immigrants spend considerably less time exposed to this type of new technology. They grew up during a less techno-frenetic era, and the current digital revolution occurred after their formative years. Many baby boomers can remember when they had only one television in the house—maybe not even color television. Some boomers find it easy to adapt to new technology—they may shop online, communicate via email, and use cell phones—but these are all conveniences they picked up as adults, after most of their brain's hard-wiring was already set in place.

Although these Immigrants are adjusting to the digital age, their approach differs greatly from that of Digital Natives. The typical Immigrant's brain was trained in completely different ways of socializing and learning, taking things step by step, and addressing one task at a time. Immigrants learn more methodically and tend to execute tasks more precisely. They are being forced to learn a new digital language—a challenge not unlike that of immigrants from other countries who arrive and do not speak the native tongue. Learning any language in adulthood requires the use of different parts of the brain than those that are used to learn a language early in life.

Techno-Phobia

The young brain's craving for new technology is just one of the forces driving the brain gap. Another equally important influence is the older generation's avoidance, and sometimes fear, of adopting new technology. Particularly when people reach their sixties and early seventies, functional impairments associated with aging may interfere with widespread use of computers and other gadgets. More than 50 percent of people age sixty-five or older have physical limitations from arthritis, or perhaps hearing or visual decline, which makes it more difficult to use current technology. Although accessible technology devices like screen enlargers and talking word processors (see Appendix 3) are available to help impaired individuals, many people don't know about these products or else don't want to bother with them.

Fear of computers creates another roadblock. For many older people, a computer or hand-held device is a black box developed for the younger generation. Their initial digital experiences were frustrating, and if they can avoid using new technologies, they often do—especially if their lives function perfectly well without them. A recent study found that older adults, compared with young people, scored significantly higher on computer anxiety ratings and were thus less likely to use computers or the Internet. Many senior executives and managers still resist even basic computer tasks like email. They have their assistants send and answer emails for them, and they prefer to review only hard-copy printouts.

A Pew Internet study found that 22 percent of Americans have never used the Internet or email and their households are not online. The vast majority of these technology-disconnects are older adults. Over the next several years, although it may still be possible to find techno-virgins, many of these technology holdouts will eventually become Digital Immigrants. We know, of course, that people of any age can overcome their techno-phobias. Studies have found that once seniors or middle-aged adults get the proper training through simple computer courses, these phobias tend to disappear.

Another technology roadblock involves how the older brain differs from that of a younger person. As with any other organ in the body, brain structure and function change as we age. Thinking and memory abilities tend to slow down because brain cells don't communicate as quickly. The levels of some neurotransmitters and neuron branches or dendrites diminish, thus decreasing neuronal efficiency in receiving messages from neighboring brain cells. For some people, it becomes more difficult to hold information temporarily in mind, a cognitive task known as working memory. This type of memory allows us to dial a telephone number immediately after hearing it from directory assistance, although we don't usually recall it a few minutes later.

Many older people are unable to process information as quickly as younger ones, so it may take them longer to retrieve names of acquaintances and co-workers. The neural circuitry of an older brain is not as poised to take on the latest fast-paced digital technology without some help. However, new research is now showing us that brain wiring can be re-fired with training: Digital Immigrants *can* catch up with Digital Natives.

Mature Brains Still Flex

Older brains may not "kick their kid's butts" at PlayStation, but their years of experience serve them well. Although the brains of Digital Immigrants may take longer to process information, mature neural circuits are often more effective in seeing the big picture, which can be optimized to improve memory and learning. If one can give meaning to information—place it into a familiar context—one will learn it more quickly and efficiently. Years of experience give us a great number of complex mental templates for storing new information: the key is to recognize those familiar templates and use them to our advantage.

Although Digital Immigrants tend to have fewer of the technology skills that sometimes appear to be second nature to Digital Natives, recent studies demonstrate that older brains *do* remain malleable and plastic throughout life. Even areas of the brain that were reserved for specialized tasks can be recruited and retrained. Using functional MRI scanning, neuroscientists studied how blind people use their visual cortex, which constitutes approximately 35 percent of total brain volume. Apparently, the brain does not waste space—the blind volunteers used their visual cortex (a brain region usually reserved for visual sensation) for increased control of their sense of touch. Since their eyes provided no incoming signals, the brain cells sought out other uses for this brain area, and in this case tactile sensation rather than visual sensation took over. This may explain why people who are handicapped in one or more of their senses, such as hearing or eyesight, often become extremely sensitive with their other senses.

In an adult, when blindness comes on quickly from retinal detachment or a traumatic injury, the brain can also retrain itself. Dr. Alvaro Pascual-Leone and colleagues at Harvard Medical School found that blindfolded volunteers used the visual cortex to process the sense of touch while learning Braille. Other research has found that the visual cortex can control other sensory functions, such as hearing.

Several studies have shown that exercising the brain with mental aerobics not only can improve cognitive performance scores but also may delay brain degeneration from diseases like Alzheimer's. A recent study of nearly three thousand older adults found that only ten sessions (one hour per week) of memory or reasoning training significantly improved cognition, and the benefits could still be measured five years after the training. Volunteers receiving the instruction also reported less trouble in carrying out everyday tasks, such as using a computer or handling their medicines.

A recent University of Michigan study found that when confronted with a constantly changing environment, older brains could turn on new regions, particularly those in the frontal lobe. Dr. Cindy Lustig and co-workers measured brain activity patterns in people eighteen to thirty years old and compared them with a group older than sixty-four. When volunteers performed easy mental tasks, their brain activity patterns were similar across age groups, but for challenging tasks, older adults recruited additional frontal brain regions not used by younger volunteers.

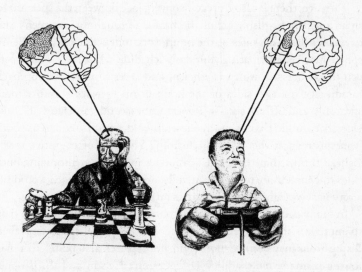

Mental aerobics or brain gymnastics may be effective for anybody's brain, regardless of age. Whether the workout involves puzzles, learning a musical instrument, reading, or playing games, the resulting benefits appear greatest when the task challenges the mind (see Chapter 8).

The Elastic Midlife Brain

The midlife brain (ages thirty-five to fifty) may be especially elastic, more so than previously appreciated. Neuroscientists say that this is the time of life when we begin to optimize all the information available to us from our experiences over the years. What may facilitate this midlife neural agility is a proliferation of the brain "glue," or glial cells (*glia* is Greek for glue)—the white matter that coats the axons that communicate between cells and continues to grow in our brains throughout middle age. Dr. George Bartzokis and colleagues at UCLA found that the quantity of this white matter coating peaks between ages forty-five and fifty, perhaps reflecting the optimal reasoning skills of people in this age group. This white matter sheathing gets more efficient at this age, and the axons can transmit information signals at an accelerated rate. In midlife, our brains graduate from a dial-up modem pace to high-speed DSL.

With age, there is also a greater convergence between the specialized right and left hemispheres of the brain. Throughout our teens and twenties, these two sides of the brain work independently—one hemisphere may be anticipating your daily schedule, while the other directs you to get dressed, which is efficient and saves time. As we approach middle age the two sides of the brain begin to work together more smoothly and effectively. Dr. Robert Cabeza and associates at Duke University found that mentally successful older adults tend to use both hemispheres together when performing a variety of cognitive tasks. Cabeza thinks that these older brainiacs may be synchronizing both sides to compensate for age-related decline, just as someone would lift a very heavy weight with both hands rather than with one.

In many people, the brain areas controlling temperament also tend to improve with age. Of course, lots of older people get cranky and their dispositions may become more rigid, but research also points to a mellowing in many older adults. Drs. Ravenna Helson and Christopher

Soto of the University of California, Berkeley, studied 123 women who were first assessed in their early twenties and interviewed again decades later. The investigators found that their likable personality traits actually peaked in their fifties and sixties, when they showed their greatest ability to remain objective, tolerate ambiguity, and effectively handle interpersonal relationships.

These kinds of age-related brain improvements may present an evolutionary advantage, which might be lost with the new generation of Digital Natives. We don't yet know how a Digital Native's brain, with its unique neural circuitry laid down early in life, will respond to age-related physiological and psychological changes. The future brain is yet to emerge.

The Graying Workforce

Thanks to advances in medical technology, people are living longer now than ever before. In the United States today, nearly eighty million people are age fifty and older. The aging of these baby boomers and their tendency to put off retirement is leading to a significant graying of the workforce. As the younger Digital Natives enter and begin to dominate the workforce, this large number of boomers staying on the job will face pressure to catch up with their younger colleagues' superior technology skills. It should be mentioned, however, that younger Digital Natives need to learn the people skills that come more naturally to their older co-workers (see Chapter 7).

The basic brain-wiring differences and experiences of Natives and Immigrants contribute to different professional expectations. Natives expect to change jobs often throughout their careers. Many Immigrants entered the workforce expecting to work for the same employer for their entire careers.

Most Digital Immigrants will continue to excel in today's workplace because of their superior interpersonal skills and greater depth of experience. Those who can master at least some new technology while maintaining their advanced interpersonal skills will be the leaders and managers of tomorrow's workforce. Not only will these business and professional achievers know how to close a deal; they'll do it efficiently, personably, and with new technology at their fingertips.

COMING TOGETHER

Bridging the brain gap will require two major interventions: We need to help Digital Natives learn to advance their interpersonal skills (Chapter 7) and teach Digital Immigrants to hone their technology skills (Chapter 8). However, both generations must maintain and enhance their abilities to talk face to face, understand subtle nonverbal cues during conversations, and build empathic ways of relating to one another, off line. Initial research and practical experience with strategies to bridge the brain gap indicate that we can have an impact on our existing neural circuitry and help ourselves to adapt to brain evolution as it unfolds.

Three

ADDICTED TO TECHNOLOGY

> *Why is it that drug addicts and computer*
> *aficionados are both called users?*
>
> Clifford Stoll,
> astronomer and author

The UPS truck pulls up in front of the house, and you wonder what Tom, the delivery guy, is bringing this time. Probably some kitchen appliance you already have two of, but this one is better! Or maybe it's those adorable boots you bought on eBay. Okay, they were a size too small, but you can have them stretched . . . You wave at Tom through the window and motion for him to leave the package at the door because you can't leave your computer just now. The eBay auction is about to end, and you *have* to get these jeans! They're *so* much cuter than the ones you have just like them in a lighter shade, and some jerk has been bidding them up all day. This is the most exciting part of the auction: you let the other bidder think they've won and then in the very last few seconds you suddenly triple their bid! You really have to pay only five dollars above the other person's last bid, but they will lose like the sad sucker they are! Three, two, one . . . It's worked! You are now the proud owner of your eleventh pair of boot-cut jeans.

Now you check the other categories you regularly bid on: three brands of shoes, four clothing labels, your good china, your everyday china, ski clothes, kid's clothes . . . There's just nothing great to bid on right now. You check your other favorite shopping websites for overstocked bedding, lamps, and furniture, but nothing piques your interest. You glance at the clock and realize you've been sitting there doing this for more than five hours. Your husband and kids will be home soon, and not only do you have to get dinner going, you have to hide that package outside— whatever it is. Hopefully there'll be some new stuff listed on eBay tonight after everyone's asleep . . .

When we think of addiction, we usually associate it with alcoholism or drug abuse. However, the same neural pathways in the brain that reinforce dependence on those substances can lead to compulsive technology behaviors that are just as addictive and potentially destructive. Almost anything that people like to do—eat, shop, have sex, gamble—has the potential for psychological and physiological dependence. But the access, anonymity, and constancy of the Internet has helped create several new forms of compulsive behavior fueled by the World Wide Web and other digital technologies.

Whether we're watching reality TV or Googling for old TV show theme songs, the brain and other organs automatically react to the video monitor's novel and staccato stimuli. The heart rate slows, the blood vessels in the brain dilate, and blood flows away from the major muscles. This physiological reaction helps the brain focus on the mental stimulus. The rapid change and flow of visual stimuli can shift our orienting response into overdrive—we continue staring at the screen but eventually we experience fatigue rather than continued mental stimulation. After a computer or television marathon, concentration abilities are diminished, and many people report a sense of depletion—as if the energy has been "sucked out of them." Despite these side effects, computers and the Internet are hard to resist, and our brains—especially young ones—can get hooked rapidly. Sales of video games are stronger than ever.

Internet addicts report feeling a pleasurable mood burst or "rush" from simply booting up their computer, let alone visiting their favorite websites—just as shopping addicts get a thrill from scanning sale ads, putting their credit cards in their wallets, and setting out on a spending spree. These feelings of euphoria, even before the actual acting out of the addiction occurs, are linked to brain chemical changes that control behaviors ranging from a seductive psychological draw to a full-blown addiction. The brain-wiring system that controls these responses involves the neurotransmitter dopamine, a brain messenger that modulates all sorts of activities involving reward, punishment, and exploration.

Dopamine is responsible for the euphoria that addicts chase, whether they get it from methamphetamine, alcohol, or Internet gambling. The addict becomes conditioned to compulsively seek, crave, and re-create the sense of elation while off line or off the drug. Whether it's knocking back a few whiskeys or whipping out the credit cards, dopamine transmits messages to the brain's pleasure centers, causing addicts to want to repeat those actions over and over again, even if the addict no longer experiences the original pleasure and is aware of negative consequences.

The mental reward stimulation of the dopamine system is a powerful pull that nonaddicts feel as well. Studies of volunteers rapt in addictive video games show that gamers continue to play despite multiple attempts to distract them. The dopamine system allows them to tolerate noise and discomfort extremely well. Previous research has shown that both eating and sexual activity drive up dopamine levels. One can only imagine the intensity of the dopamine bursts produced from an interactive video game with a sexual theme (and they're out there).

As Internet addiction takes hold, the brain's executive region, known as the anterior cingulate, loses ground. This is an area in the front part of the brain that is responsible for decision making and judgment. Intervention for these addictions involves not only holding the dopamine system at bay but also strengthening these anterior neural circuits.

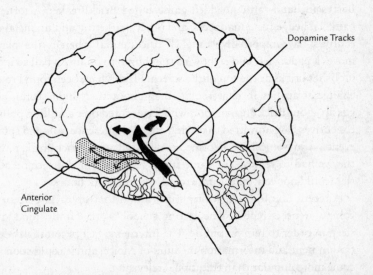

ANYONE CAN GET HOOKED

Internet addiction afflicts people from all walks of life: homemakers in their thirties and forties, teenagers, business people in their fifties and older, college students, and even kids under ten. Everyone is at risk to get hooked on Web applications. In February 2007, the *Los Angeles Times* reported that after working nineteen stellar years for a large computer company, a man was fired for visiting sexual chat rooms during his breaks. The man, married with two children, claimed that the chat room visits helped ease the stress he had continued to experience since the Vietnam War. At the time of the news report, he was suing the company for wrongful termination.

Many kids and teens may not exactly be addicted, but the pull of new technology can cloud their judgment. The anterior cingulate in their brains often loses out to the dopamine rush they get from text messaging with their friends. Teenagers' text messaging while driving may cause thousands of fatal car accidents across the United States. Although text messaging is much more distracting than merely talking on a cell phone, as of July 2007, only a handful of states in this country had outlawed text messaging while driving.

Business executives have found another reason to keep their Black-Berrys in hand—an embedded game called BrickBreaker. Attorneys, bankers, hedge fund managers, and other marketing and financial executives have reportedly "caught" Brickmania, wherein the player moves a paddle left and right with the thumb to bounce a ball so it demolishes bricks at the top of the screen. Players have been found to exchange strategies in chat rooms, brag about their high scores, and gossip about BrickBreaker idols who have scored over a million points. Executives have admitted to playing during conference calls and sports events, and some have become so obsessed with BrickBreaker that they've had to remove the game from their hand-held device because they could not control the urge to play during work hours.

A recent Stanford University study found that up to 14 percent of computer users reported neglecting school, work, family, food, and sleep in order to remain online. The Internet is fast becoming the entertainment and information medium of choice, and it could soon become more popular than traditional television.

It has been reported that college students with difficulties adjusting to campus life may use the Internet to escape everyday stress. Rather than face the challenges of face-to-face social life, they feel a greater sense of control by using social networking sites, email, instant messages, and chat rooms. More than 18 percent of college students are pathological Internet users, and 58 percent report that their excessive Internet use has disrupted their studying and classroom attendance and also lowered their grade point average.

It is not the Internet itself that is addictive, but rather the specific application of choice. People can get hooked on database searching, online dating, Web shopping, porn sites, or even checking their email. Others become compulsive online gamblers, stock traders, gamers, or instant message senders.

Even if you are not addicted to the Internet or a related technology, you may be struggling with its enticement, perhaps letting it get the better of you once in a while. If you're curious as to whether you have a problem, take the Technology Addiction Questionnaire in Chapter 6.

Part of the appeal of new technology is the sense of control it gives us. The direct and instant command we have over our computers empowers us. We have the power to turn our computers on and off whenever we feel like it; we can make them stand by, hibernate, or reboot. We can gauge the pace of our communications, or not communicate at all if we wish.

But for individuals at risk for addiction, the computer and Internet can provide a false sense of control. The screen, keyboard, and mouse become extensions of the individual—a hardware/software link to the global, Internet-connected world. Compulsive users report feeling a sense of liberation and anonymity online, so they often say or write things they might not otherwise reveal about their personal lives. Some users get a thrill from making up completely false personalities. What some Internet users don't realize is that once you put your thoughts and feelings into words on the Web, they are public forever—and accessible not just to friends and family members but also to work associates, job recruiters, and people without your best interests in mind. With the growing interest in weblogging, supervisors are beginning to keep track of their employees' blogs and are not hesitating to fire workers who post blogs that are perceived to compromise the company's information or brand.

A Seattle contractor who worked for Microsoft noticed several Apple Power Mac G5 computers sitting on one of Microsoft's loading docks. Even though Microsoft makes software for its rival Mac operating system, and routinely tests competing technologies, this worker thought it amusing to see these Mac computers arriving on the Microsoft campus. He snapped a photo and posted it on his weblog with this caption: "It looks like somebody over in Microsoft land is getting some new toys." He was fired the next day.

Whether it's blogging or Internet shopping, addictive behaviors get people into all kinds of trouble, not just job-related trouble. Addiction doesn't happen overnight—habit-forming behavior patterns build gradually. Usually, an individual begins Internet use casually, but eventually the emotional charge and the amount of time spent online progress, and the brain needs a stronger dopamine boost. Soon a psychological dependence on the Internet develops, causing the person to experience discomfort when not online, while at the same time the person may be developing tolerance to the effects of being online. The user may then feel the need for more time online or possibly more exciting online sites. Pathological Internet users typically start out in complete denial that they have any problem controlling their online activity. Although the amount of time spent on the computer is usually considerable, the disorder has as much to do with whether the behavior interferes with the person's everyday life by disrupting his or her job, family life, or social activities.

The driving force of addiction depends on the individual. Genetics plays a role. Some people inherit a tendency to get hooked on almost anything, and the Internet supports various forms of addictive behavior that often occur off line as well, such as gambling, eating, sex, and shopping. Others are looking for an escape from depression, anxiety, boredom, or interpersonal conflict. Peer pressure spurs many young people to get hooked on interactive online activities involving chat rooms, social networks, or virtual games.

Addiction experts have proposed criteria for Internet addiction disorder, which include mood changes, tolerance, withdrawal symptoms, and relapse (see Sidebar). Some experts estimate that 10 percent of

Internet users meet these criteria for addiction, which are similar to those for gambling and shopping addictions, wherein the user is addicted to a process rather than a substance such as drugs, alcohol, tobacco, or food. However, whereas a substance abuser strives for complete abstinence (except for food), an Internet addict more often tries to attain moderation. Recently, the American Medical Association recommended additional study to determine whether video gaming and Internet addictions should be considered official diagnostic categories.

PROPOSED CRITERIA FOR INTERNET ADDICTION DISORDER

The following specific criteria must be present:
- *Preoccupation.* The individual thinks about previous online activity or constantly anticipates the next online session.
- *Tolerance.* Longer periods online are needed to feel satisfied.
- *Lack of control.* The person is unable to cut back or stop online activities.
- *Withdrawal.* Attempts to decrease or stop Internet use leads to restlessness, irritability, and other mood changes.
- *Staying online.* The user repeatedly remains online longer than originally intended.

In addition, at least one of the following criteria must be present as well:
- *Risk of functional impairment.* Internet use has jeopardized the loss of a job, educational or career opportunity, or important relationship.
- *Concealment.* The user lies to others in order to hide their Internet activities.
- *Escape.* The individual goes online to relieve uncomfortable feelings, escape problems, or not deal with personal relationships.

Internet addicts typically spend forty or more hours each week online in addition to online work time. If you calculate the number of hours needed to eat, work, travel, dress, and bathe, that leaves approximately four to five hours a night for sleep before a person hops back on the computer. Most addicts lie about it to others and tend to get defensive when family and friends question the amount of time they spend online. These addicts routinely experience apathy, depression, anxiety, restlessness, fatigue, irritability, and clouded thinking.

Internet addiction has physical side effects as well. Too much time at a computer display terminal can lead to muscle strain, eye discomfort, and headaches. Letters appear less precise and not as sharply defined when viewed on a video screen rather than a printed page. Screen contrast levels are often less than optimal; moreover, reflections and glare from the screen can make it even harder to see. Repetitive mouse use can cause tendinitis and muscle cramping in the hands and upper arms. Because vision and joint mobility generally decline with age, older people are more sensitive to these kinds of physical symptoms, although with continual use they can occur at any age. One recent study found that more than two hours of daily computer use was significantly associated with neck, shoulder, and low back pain in adolescents.

EMAIL JUNKIES

Now that so many of us are continually logged on to the Web, lots of people no longer sit and read a newspaper or a magazine but instead get their news from the Internet as they furiously try to keep up with their email and other online activities. Many use hand-held devices to surreptitiously check their email during business meetings, corporate retreats, their kid's soccer games, and even church services. There are CEOs of Fortune 500 companies who check their BlackBerrys after every golf shot, and some people actually refuse to vacation anywhere they cannot get a high-speed Internet connection to their email and other Web-based sites at all times.

Part of what makes email so addictive is that it follows the rules of operant conditioning, which means that the behavior is shaped by its consequences (see Box). When you check email, you get intermittent positive responses. Sometimes you receive good news: the arrival of an old friend, perhaps a great joke, or a long-awaited response to a request. Occasionally, you receive fantastic news, such as word that your lost winning lottery ticket was found at the dry cleaners. But more often, a neutral, boring, or distressing mail notice or spam gets through. You can never tell in advance whether checking your email will be pleasurable or not, so you keep on checking, checking, and checking. Behavioral psychologists have detailed how the principles of reward and punishment reinforce this behavior, and they have found that using consistent rewards—good news all the time—is less motivational than randomly occurring rewards. Much like gambling addicts, people keep up the behavior because "next time" may bring the big payoff. The brain's neural circuitry is prewired for this response.

EMAIL AS AN EXERCISE IN OPERANT CONDITIONING

How One Gets Hooked

Let's say you open your first few emails, and the messages are negative (unwanted spam, annoying jokes or chain letters from so-called friends), or perhaps a reminder of work you have been avoiding. You may feel like giving up email altogether. The faces below represent the negative emotional neural networks tweaked by these emails:

☹☹☹☹

Then suddenly you get an email that thrills you—a big raise at work or a note from your wife that your son got straight A's.

☹☹☹☺

That happy face email will excite an entirely different cluster of neural circuits, causing dopamine to surge through your brain. That consequence reinforces future behavior to check email. Now you are probably willing to open many more emails, hoping for a future happy face.

☹☹☹☹☹☹☹☹☹☹☹☹☹☹☺

Operant conditioning, wherein the consequence of a behavior reinforces future behaviors, is a very powerful mechanism. It drives addictions and compulsive behaviors. Consider whether all your emails elicited a happy face neural network:

☺☺☺☺☺☺☺☺

Email would not have the same charge—it would be no different from stepping into a nice warm shower or taking money out of an ATM. It would be a positive experience each time (unless, of course, you have no money in your account).

Despite the addictive potential of email, help for email junkies is available. Addiction experts and treatment centers are beginning to offer programs for a variety of technology addictions (see Chapter 7 and Appendix 3). For example, an executive coach in Pennsylvania has actually devised a twelve-step program designed to tackle email addiction, which some users say can become an obsession so great that it diminishes their productivity and consistently intrudes on their normal daily life.

VIRTUAL GAMING—BET YOU CAN'T PLAY JUST ONE

His skateboard lay on the side of the house, along with his bike, basketball, and soccer gear. Eleven-year-old Ryan hadn't touched any of it in weeks, maybe months, ever since he'd started playing The Game. He'd race home from school, do a perfunctory job on his homework, and run upstairs to the computer, where he would transform into his Game identity: Swordsman of Farlander, Protector of the Grand Vision. Many of his friends were online there too, all in their various identities, but Ryan was the god of them all—he had reached Level 10—and no one else, no matter how many hours they had spent playing The Game, had reached Level 10 yet. Of course, Ryan, or rather the Swordsman, had had to kill quite a few of his friends, capture their treasures, and steal their visions, but that's how you got ahead in The Game.

During a particularly gory battle with his best friend Dylan's alter ego Titanus, King of the Mountains, Ryan's mother came up and said it was dinnertime. Ryan barely acknowledged her; he was so wrapped up in the battle at hand. Killing Titanus would reap a hefty treasure and several visions. The fact that it would devastate his best friend and send him back to Level 1 was irrelevant.

Ryan's Mom, hating this ridiculous video game and herself for buying it, repeated, "It's time for dinner. Did you hear me, Ryan?" "Yeah, okay, whatever," he responded without taking his eyes off the screen or his fingers off the keyboard. Mom: "Maybe you don't understand me. I mean NOW." Ryan: "Okay. Right after I kill this guy." Mom: "No, not after you kill this guy. Now." She reached down and shut off the computer. Ryan shrieked, "Mom! What did you do?! I didn't save my game! I'll have to go back to Level 1!" Mom: "I have a better idea. Why don't you go back to being a normal kid? This game is going in the trash." As she removed the game's CD, required for online play, she kissed Ryan's cheek and said, "Wash your hands, honey, we're having roast chicken." "I hate chicken!" Ryan wailed, "I'm the Swordsman!" Mom smiled, "Great. You can slice the chicken."

Television used to be the medium of choice for relaxing and tuning out, but today young people in particular are opting for computer or video games. The Interactive Digital Software Association reported that in 2006 approximately 145 million people, or 60 percent of Americans, were playing video or computer games. While females are more likely to stay in touch with friends through social networks like MySpace and Facebook, chat rooms, and instant messaging, males tend to be more comfortable with virtual gaming social networks: 80 percent of online virtual gamers are young men—and not just teenagers; the average age is twenty-eight.

Web domains supporting games are referred to as sticky sites, since online gamers tend to stick to them. In a study of the Internet heroic fantasy game Everquest, players spent an average of twenty-two hours each week gaming, and some of them logged on more than eighty hours a week.

Players explore three-dimensional environments while interacting with friends or thousands of other players in real time. They race cars, conquer sci-fi future worlds, or sometimes become lost in historic or futuristic fantasylands. Many assume identities they create for themselves, and "level up" while collecting valuables and weapons. Often, game addicts escape into these cyber worlds in order to feel engaged and powerful, while off line they may feel lonely and listless.

For some, online multiplayer games become an intense form of fantasy social networking. The characters and relationships that people take on in these games can begin to replace their real-life relationships. Recently, it was reported that a fifty-three-year-old man was playing the game Second Life for as long as fourteen hours a day. In his virtual world, instead of his regular job as a call center operator, he had re-created himself as a successful entrepreneur. While playing the game, his character met a female character, who was created and controlled by a real woman also playing the game. Their virtual relationship flourished, and they had a wedding within the game. The man claimed that he "truly cared for her," but had no plans to meet the real woman outside the game. His real-life wife complained that the game distracted him from his family and job and that they had little or no intimacy. He said that his real wife was just jealous of his virtual wife.

Besides seducing people into a virtual life and suppressing frontal lobe function in teenagers (see Chapter 2), these games have other brain effects that fuel their addictive potential. During a game, dopamine is released into the player's brain, causing intense pleasure and a sensation of being in control. Gamers report additional enjoyment when the game or online interaction provides a virtual social network.

ONLINE PORN OBSESSION

Sexually explicit images and content are easy to access on the Internet unless some type of parental or adult control is installed. Although only 4 percent of websites display sexually related material, at least one-third of Internet users engage in some type of online sexual activity. Forty million Americans visit Internet porn sites at least once each month, and 35 percent of all downloads are pornographic. Some people simply email sexually explicit jokes or humorous images; others find the accessibility, affordability, and anonymity of more intense cyber sex irresistible.

Though pornographic material does not dominate cyberspace, it is certainly well represented: Enter the word "sex" in a search engine, and you'll be directed to over four hundred million pages. Dr. Amanda Spink and her associates at Pennsylvania State University analyzed patterns of Web searching for sexual information. They found that sexually related search sessions were usually longer and contained more queries, compared with nonsexual sessions: nearly 40 percent of sexually related query sessions lasted longer than six minutes, compared with only 22 percent of nonsexual queries that continued for that length. Sexual queries also involved simple keyword searches using such generic terms as nude, sex, and naked, whereas nonsexual queries used more complex and varied language.

Although some people engage in cyber sex within the context of an existing or intimate relationship (e.g., lovers geographically separated), it can lead to problems, especially when it involves strangers or becomes repetitive or compulsive. For individuals genetically or otherwise predisposed to addictions, viewing pornographic computer images or sending and receiving sexually explicit messages can quickly turn into a habit. The accessibility and anonymity of the Web, both at home and at work, help make it particularly tough for sex addicts to abstain.

Some people get hooked on viewing seductive photos and film clips; others find that eventually they need more than just cybersexual relationships to satisfy their addiction. Sometimes the Internet becomes just one aspect of an addict's sexually obsessive behaviors, which can include a variety of off-line compulsive activities.

Sexually explicit visual images trigger dopamine brain messages that are nearly identical to the neural-chemical transmissions in cocaine or heroin addicts when they get their drugs. These individuals may spend hours into the night searching for just the right image that triggers the dopamine reward, and their prefrontal cortex organizes the search, storage, and retrieval process necessary to keep feeding that needy dopamine system.

Because over 70 percent of companies provide Internet access for their employees, cyber sex has become a workplace problem for many businesses. A 2006 study of more than thirty-four hundred volunteers found that such Internet accessibility predicted online sexual activity at work, and the more time spent at sex sites, the lower the productivity at work. One corporate study found that 41 percent of the employees who were reprimanded for computer abuse at work were using the computer for pornographic purposes.

LAS VEGAS AT YOUR FINGERTIPS

Gambling has always posed an addiction problem for some people. It fits a perfect operant conditioning model, with its intermittent rewards system. The occasional big payoff or reward reinforces the perpetual risk taking of gambling. Pathological gamblers often devastate their lives and those of the people around them with their financial and emotional problems. Approximately one out of four members of Gamblers Anonymous reports having lost jobs, families, or both because of gambling habits.

Technology has played a role in gambling for many years, but yesterday's one-armed bandits have been replaced by today's Internet betting sites. Illegal Internet gambling, available twenty-four hours a day, seven days a week, is on the rise. A 2005 survey found that 4 percent of Americans gamble online. Nearly 40 percent had started gambling on the Web in the past year, and 70 percent began during the previous two years.

When players bet online with electronic dollars, their perception of the value of money diminishes, and they run the risk of excessive gambling and debt accumulation. Gambling websites do not always verify whether the gambler is of legal age. Sometimes dishonest offshore Internet gambling businesses completely remove sites within minutes in order to steal credit card numbers, and there is minimal regulatory control. Some of these sites have embedded keywords like "compulsive gambling," so recovering gamblers searching the Web for information on how to kick their online habit may get automatically redirected to actual gambling sites. New laws may help to curtail online betting, and officials in the United States have stepped up their enforcement of existing legislation. In 2006, Congress passed the Unlawful Internet Gambling Enforcement Act, which bars credit card companies and banks from assisting people with fund transfers to online casinos.

People who use the Internet to gamble are likely to have more serious gambling problems than those who visit casinos. Drs. George Ladd and Nancy Petry of the University of Connecticut found that in a sample of nearly four hundred individuals, only 8 percent were Internet gamblers, but 74 percent of these Internet gamblers had pathological gambling problems, compared with only 22 percent of the off-line gamblers.

Research shows that dopamine, the brain messenger that mediates pleasure and reward-seeking behavior, also contributes to Internet gambling. Doctors from the Mayo Clinic and other medical centers recently reported on Parkinson's patients who were taking drugs that mimic the effects of dopamine. These drugs helped relieve the tremors and stiffness of Parkinson's disease. However, many of the patients in the study developed Internet gambling problems, compulsive online sex issues, and Web-shopping disorders. Fortunately, the pathological gambling and other behaviors disappeared when the drug was stopped or the dose was lowered.

SHOP TILL YOU DROP

Thanks to the phenomenal growth in e-commerce, it seems that almost everyone is shopping online. Many people start out reluctant to share their credit card numbers with anonymous websites, but eventually they are made to feel more comfortable with PayPal and other electronic commerce payment systems, and they begin to prefer the convenience of online shopping. No need to search for a parking space, drag around large bags, or even leave your computer station—just search the keywords for the item you want, compare prices, and punch in your credit card number. You can even customize your purchase, whether it's a new car, a used refrigerator, or an engraved iPod.

Unfortunately, online shopping can easily become a compulsive pursuit. Since no money changes hands, and no cash register rings up a total, many shopping addicts don't feel that they're really spending. Online shopping becomes addictive for the very same reasons that off-line shopping does. The instant thrill of making the purchase, supported by the pleasure of anticipating and acquiring the item, can become too much to resist.

Buying through an auction can shoot even more dopamine into the brain and become even more addictive. The delight of bidding and beating out all other bidders can be thrilling for those hooked on eBay and other auction sites. The actual item purchased may be less important than the ritualized and repetitive behavior patterns that produce these momentary pleasures. As with any addictive behavior, people usually don't realize it's a problem until it starts interfering with other areas of their life—financially, professionally, or personally.

GETTING HELP

Whether you are truly addicted or are just having trouble saying "no" to some websites, help is available. Experts agree that kicking habitual behaviors that are reinforced by the brain's dopamine reward system usually involves a combination of approaches. For severe addictions, psychotherapy, twelve-step programs, and support from family and friends can be effective, but the addict must be motivated to give up the addiction. The bottom line is that no matter how strong the support system, a person must truly want to quit an unhealthy habit.

In China, where an estimated two million youths are addicted to the Internet, a boot camp for Web addicts has successfully treated thousands. Using tough-love strategies and daily physical exercise, the government-funded Internet Addiction Treatment Center helps addicts strengthen their sense of belonging to a group. Numerous other addiction treatment centers and programs are beginning to offer help for Internet addiction throughout the United States and Europe.

After getting a sense of whether or not the Internet, video gaming, or other technologies may have become addictions for you or someone you know (Chapter 6), you'll find specific strategies for helping people wean themselves from technology addictions in Chapter 7. These techniques can also help those who are not truly addicted but find themselves spending too much time on the Internet. In addition, other exercises for strengthening social skills and reconnecting face to face can help people not only enjoy more fulfilling time off line but also live more balanced and satisfying lives.

Four

TECHNOLOGY
AND BEHAVIOR:

ADHD, Indigo Children, and Beyond

> *To err is human, but to really foul things up*
> *you need a computer.*
>
> Paul Ehrlich

After dinner, Rita took a break from finishing her email correspondence to watch the evening news. While she gazed at the journalist recapping the headlines of the hour, her eye followed the crawler at the bottom of the screen, and she split her attention to decide whether the crawler news was more important than the hour's headline. A moment later, a tiny celebrity popped up in the lower right-hand corner of the screen, along with graphics advertising an upcoming primetime drama. The TV was bombarding her with so many different types of information that she was getting a headache. Thankfully, a commercial interrupted Rita's stressful TV watching, and she took a much needed break from her break.

Rita checked on her thirteen-year-old daughter's homework progress. In her room, her daughter was seated in what Rita affectionately called the cockpit—the chair at her desk surrounded by computer paraphernalia and books, with her keyboard in her lap, her iPod headphones in her ears, video chatting with her girlfriend, which she told Rita is their way of "studying together," while she checked Wikipedia for definitions, and uploaded new images onto her MySpace page. Rita asked how her history homework was going. She gave Rita a thumbs-up before going back to her video chatting. If she weren't bringing home

(continued)

(continued)

A's, Rita would have to put her foot down. This high-tech craziness was getting out of hand. As Rita left her daughter's room she longed to really kick back—maybe crawl into bed and read a novel—but her cell phone was ringing, she could hear a fax coming in, and she needed to respond to a few more emails.

New technology bombards us from everywhere—digital billboards, supermarkets, computer screens, and, of course, our cell phones. A rather startling consequence of the perpetual digital stimulation of our brains and the multitasking it seems to force upon us is its effect on our ability to focus attention on any one particular thing.

Digital Immigrants over forty tend to have a harder time dividing their attention and keeping pace with this new frenetic and fragmented environment. However, the brain's neural circuitry does have the capacity to create new pathways to process even this hyperactive, attention-splitting data input. As a result, we are developing alternative ways to learn and think. In order to adapt, our brains are learning to access and process information more rapidly and also to shift attention quickly from one task to the next.

You can observe this process whenever you try out an unfamiliar computer program. Initially, your mind struggles to understand and manipulate the program. You may wade through it awkwardly, making mistakes, but with practice it gets easier. Then, before you know it, you've become proficient in the program, and soon you're an expert.

DRIVEN TO DISTRACTION

A certain level of brain stimulation is healthy and enjoyable, but when exposure to new digital technology becomes excessive, the brain response can become maladaptive, especially if someone carries a genetic risk. Some individuals cannot effectively handle the multitasking demands of modern technology, and sometimes syndromes such as attention deficit disorder (ADD) or attention deficit hyperactivity disorder (ADHD) can result (see Sidebar).

DIAGNOSTIC CRITERIA FOR ADD/ADHD

The following features must be present:

I   Either *inattention symptoms* or *hyperactivity-impulsivity symptoms* that are maladaptive and present for six or more months:

A   Six or more of the following *inattention symptoms*, which occur frequently:

i   Failure to pay close attention to details or careless mistakes

ii   Difficulty sustaining attention

iii   Does not seem to listen when spoken to directly

iv   Does not follow through on instructions

v   Difficulty organizing tasks and activities

vi   Avoids/dislikes tasks requiring sustained mental effort

vii Loses items necessary for tasks or activities

viii Easily distracted by extraneous stimuli

ix   Forgetful in daily activities

B   Six or more of the following *hyperactivity-impulsivity symptoms*, which must occur frequently:

i   Fidgets or squirms in seat

ii   Gets up from seat when sitting is expected.

iii   Inappropriately moves about, or has feelings of restlessness

iv   Difficulty playing or engaging in leisure activities quietly

v   Always "on the go" or "driven by a motor"

vi   Talks excessively

vii Blurts out answers before questions have been completed

viii Difficulty waiting turn

ix   Interrupts or intrudes on others

II   Shows some symptoms before age seven

III Shows some impairment in two or more settings

IV Significant impairment in work, school, or social life

Adapted from *Diagnostic and Statistical Manual
for Mental Disorders (2000)*

An estimated 5 percent of children in the United States have ADHD, suggesting an increase in the diagnosis in recent years. This trend may reflect a true rise in incidence or, perhaps, a greater recognition of the condition. Both genetics and environment are believed to play a role in the cause. The brains of young children are sensitive to visual and auditory stimulation, which shape early neuronal development and synapse growth. Chronic exposure to such technology as TV, videos, and computers, contributes to the risk of ADHD, particularly when that exposure occurs in the first few years of life.

Drs. Philip Chan and Terry Rabinowitz of Brown University assessed the amount of time that ninth- and tenth-graders spent using the Internet, watching television, and playing video games. The investigators found that adolescents who played console or Internet video games for more than one hour each day had greater symptoms of ADHD or inattention than those who did not. More gaming time also increased their risk for school problems.

Other studies have shown that Internet addiction in elementary school children significantly increases the likelihood of ADHD and inattention symptoms. In a 2007 study of more than two thousand students, investigators from Kaohsiung Medical University in Taiwan found that Internet addiction was associated with a significantly higher rate of ADHD. Psychiatric investigators in South Korea also found that 20 percent of Internet-addicted children and teens demonstrated relatively severe ADHD symptoms.

In similar studies of television exposure, Dr. Dimitri Christakis and colleagues at the University of Washington evaluated more than thirteen hundred children and found that 10 percent of them had problems with attention. On average, a one-year-old child spent 2 hours each day watching television, and by age three, the daily average nearly doubled to 3.6 hours. The study showed that by age seven, the more a child watched television each day, the higher the risk for a diagnosis of ADHD. This large-scale longitudinal study definitely confirmed the suspected association between excessive television viewing and ADHD, but experts are not sure what aspect of television viewing may be causing the negative brain effects. One theory is that the rapidity of TV image changes leads to sudden shifting among multiple neuronal circuits. When this occurs over an extended time in a young, developing brain, the normal laying down of neuronal pathways is disrupted, which may lead to impaired attention abilities.

National professional organizations like the American Academy of Pediatrics have responded to such scientific evidence by warning parents to limit their child's television watching; they actually recommend zero television watching in children younger than two years of age. But the scientific evidence and the recommendations that toddlers watch little or no TV have not been enough: one in five American children under age two has a television in his or her room, and one-third of kids ages three to six have TV in their rooms. Most families explain that this allows them to watch their own shows elsewhere in the house.

MULTITASKING BRAINS

While toddlers need to say no to TV, adults must learn to better manage their own arsenal of smart phones, organizers, and electronic gadgets. Many of us are challenged in the face of persistent information overload and fast-paced work schedules. We find ourselves multitasking nonstop just to keep up.

Many baby boomers complaining of distractibility, impulsivity, and hyperactivity are embracing a hot new diagnosis: adult ADHD. They have an armory of medical and psychotherapeutic treatments to thwart their struggles with time management and organization. Adults with ADHD have a harder time moving from one task to the next. They find themselves unable to finish an activity when frequently interrupted. If tasks are completely different, they cannot multitask at all.

Today, the rapid pace of information assaulting our brains challenges our ability to pay full attention to any one thing. Radio and television announcers speak in time-compressed sentences. Our laptops, fax machines, and instant messages pressure us into quick responses that force us to sacrifice detail and accuracy. Many people are replacing depth and subtlety in their thinking with quick mental facts that may only skim the surface. The clutter, noise, and frequent interruptions that assail us further fuel this frenetic cognitive style.

Some professionals argue that ADHD is not truly a diagnostic disorder but rather the brain's adaptation to its perpetual exposure to multiple bits of information delivered through today's fast-paced technology.

They contend that ADHD is not an illness but simply the result of new wiring patterns for the modern brain as it adapts to ever-present technology. Eventually these adaptations may redefine mainstream culture.

Though we think we can get more done when we divide our attention and multitask, we are not necessarily more efficient. Studies show that when our brains switch back and forth from one task to another, our neural circuits take a small break in between. This is a time-consuming process that reduces efficiency. It's not unlike closing down one computer program and booting up another—it takes a few moments to shut down and start up. With each attention shift, the frontal lobe executive centers must activate different neural circuits.

Neuroscientists have found that we lose time during these switches, especially when a mental task is new or unfamiliar. Psychologist David Meyer and colleagues at the University of Michigan studied brain efficiency when volunteers quickly switch their mental workouts from identifying shapes to solving math problems. Both tasks take longer, and mental accuracy declines, when the volunteers are required to make attention shifts, compared with when they focus on only one task for an extended period. Switching back and forth between the two tasks, like answering email while writing a memo, may decrease brain efficiency by as much as 50 percent, compared with separately completing one task before starting another one.

Dr. Gloria Mark and associates at the University of California at Irvine studied the work habits of high-tech office employees and found that each worker spent an average of only eleven minutes per project. Every time a worker was distracted from a task, it took twenty-five minutes to return to it. Such distractions and interruptions not only plague our work environments but also intrude on our leisure and family time. The bottom line is that the brain seems to work better when implementing a single sustained task than when multitasking, despite most people's perception that they are doing more and at a faster pace when they multitask.

Studies of gender differences in multitasking generally find that women perform better on verbal tasks (left brain), while men excel at spatial tasks (right brain). In addition to gender, various factors influence our ability to multitask, including the type of tasks and their level of difficulty. Some combinations of tasks actually appear to improve mental efficiency. Many people notice improved cognitive abilities while also listening to music. Neuroscientists have found that some surgeons perform stressful nonsurgical tasks more quickly and with increased accuracy when listening to their preferred musical selections. Music appears to enhance the efficiency of those who work with their hands. Music and manual tasks activate completely different parts of the brain; thus, effective multitasking sometimes appears to involve disparate brain regions. However, if you are working while listening to music you do not like, it may be distracting and may decrease the efficiency of your multitasking.

Multitasking has become a necessary skill of modern life, but we need to acknowledge the challenges and adapt accordingly. Several strategies can help, such as striving to stay on one task longer, and avoiding task switching whenever possible. We can also learn and build multitasking skills with practice (see Chapter 7).

INDIGO CHILDREN

Brain evolution sometimes results in accidents of nature that help a species adapt and move toward a higher evolutionary level. For example, when our ancestors learned to use tools, they not only became more efficient hunters but developed manual dexterity and language (see Chapter 1). Such evolutionary side-steps—the consequence of our evolving brains and their ensuing new neural pathways—can also result in what seem to be unexplained behavior clusters.

One example of an apparently new and unusual behavior cluster has been termed the Indigo Children. Named for the deep blue auras that some psychics claim to see around such children, the concept of Indigo Children grew out of the New Age movement. Popularized in websites, books, films, and television, these children reportedly have extraordinary creative, psychic, or healing powers.

Descriptions of Indigo Children note several clusters of behavior:

- Exceptional creativity, intelligence, intuition, empathy, and/or abstract thinking

- Little interest in traditional schooling

- Frustration or boredom with assigned tasks, rituals, systems (e.g., waiting in line), or any activity not involving creative thinking

- Elevated self-esteem

- Difficulty with authority and frequent refusal to follow directions or orders

- Antisocial behavior and a sense of entitlement

- Difficulties getting along at school

- Tendency to turn inward and frequently daydream

Despite speculation about the Indigo Children's paranormal attributes and spiritual abilities, most of our knowledge of these kids is anecdotal and lacks scientific backing. Although these children may be highly intelligent, they may also have learning disabilities and possibly suffer from ADHD, autism, or a related condition known as Asperger's syndrome, characterized by high intellectual capacity but impaired social relationships.

The true cause of the Indigo behavior cluster is not known. Many of the attributes are consistent with an ADHD diagnosis, while other behaviors are typical of gifted children—those with extraordinary talents, maturity, intuition, or creativity. Thus, Indigo Children may represent a group of highly intelligent children with coexisting attention deficit disorders.

Because of their exceptional intellectual abilities, extremely gifted children often do not fit in with the mainstream. They find school boring and get impatient with the slow pace of routine curricula. They may space out during school or at home and are often oppositional and argumentative. Many become underachievers and suffer from typical ADHD symptoms: impulsivity, fidgeting, disorganization, absentmindedness, and poor attention to detail.

Some gifted children with undiagnosed learning disabilities have trouble getting along with other kids and appear disorganized and distractible. Some of these children have selected areas of mental agility but impairment of other cognitive skills. For example, a child may have relative weaknesses in sequencing and phonetics but outstanding reasoning and visual-spatial abilities.

The exact proportion of ADHD children who are intellectually gifted is not known. Studies have found that as many as a third of ADHD children score in the 90th percentile in a standardized measure of creative thinking, but medicines to treat their symptoms could have influenced the results. Studies that take into account the effects of ADHD medicines find lower rates of exceptional intellectual abilities.

Brain imaging studies of highly intelligent children have found wiring and brain maturation patterns that differ from those of other kids. Dr. Philip Shaw and colleagues at the National Institute of Mental Health studied 307 children over a seventeen-year period and found that patterns of brain development differed according to IQ High-IQ children (those who scored between 121 and 149) showed maximal thickening of the outer layer of brain cells (i.e., the cortex) at around age thirteen, whereas cortical thickening in average kids (IQ scores between 83 and 108) peaked at age seven. After peak thickening, the brain neural circuits become more refined during a pruning process that continues through adulthood (see Chapter 1). Experts speculate that this more lengthy cortical thickening and thinning pattern reflects a fine-tuning process occurring in the more plastic and malleable neural circuitry of gifted children. These differences result partly from genetics, but nongenetic factors, such as diet, education, and other influences, are important as well.

Brain imaging studies of Indigo Children would help sort out their underlying neural circuitry. And, as with many behavior symptom clusters, the underlying causes could surely vary from case to case. One explanation for the Indigo phenomenon is that the altered neural circuitry resulting from early and constant digital and video technology exposure not only contributes to attention problems in today's young people but also heightens some forms of creativity and insight. What some describe as Indigo Children, or perhaps gifted ADHD kids, may in part be a consequence of the digital age and our evolving brains.

CAN TV TRIGGER AUTISM?

Cornell University economist Michael Waldman became interested in possible causes of autism after his two-year-old son was diagnosed with autism spectrum disorder. Waldman had noticed that his son had begun watching more television the summer before the diagnosis, when his baby sister was born, so Waldman decided to sharply restrict the boy's television watching. The family noted improvement in the next six months, and the boy eventually recovered completely.

In order to more systematically explore a possible TV/autism connection, Waldman used a method familiar to economists to sort out the cause and effect of two variables of interest. Rather than wait for study results from a systematic randomized clinical trial, Waldman identified what economists call an instrumental variable: some random or natural event that correlates with one variable but has no effect on the other. In this study, the third variable was rainfall.

Waldman and colleagues first studied weather data and television-viewing patterns in children and found that when it rains or snows, kids on average spend more time in front of the TV set. They then chose three states that have variable precipitation patterns from year to year—California, Oregon, and Washington—and found that children growing up during periods of higher precipitation had a greater likelihood of being diagnosed with autism. To further check on whether their hypothesized connection made sense, they looked at another variable associated with TV watching—subscriptions to cable—and found higher rates of autism among households with cable television in those three states.

Their findings stirred up controversy in a field that has seen previous cause-and-effect theories debunked. In the 1940s, psychologist Bruno Bettelheim hypothesized that emotional withholding by mothers caused autism, but large-scale studies have since found no such connection. Genetic causes have been documented, but as with any complex disorder or disease, environmental factors play a role, and a wide range of conditions can be diagnosed in the autism spectrum. In some cases, genetic factors play a greater role; in others, environmental influences may contribute more. Waldman's position is that until scientists can definitively sort out these potential causes, parents might wish to follow the American Academy of Pediatrics recommendations to restrict television watching by their young children.

Autism has been found to involve not only language and social problems but also widespread brain changes that alter many aspects of thinking and behavior, including sensory and motor skills, attention, and problem solving. When autistic research subjects are given a face-recognition task during functional MRI scanning, they are more likely than nonautistic volunteers to process the information without using much of the neural circuitry in their frontal lobes. They also tend to analyze facial features more as objects than as components of a human being. Thus, many scientists suspect that autism may result from the brain's inability to integrate complex information coming from various parts of the brain.

A common feature of autism is a reluctance to interact with other people. These individuals usually have a hard time making eye contact and participating in face-to-face interactions. For most of us, direct eye contact can convey a sense of intimacy or threat, which autistic individuals have difficulty tolerating. In brain imaging research addressing these issues, neuroscientists at the University of Wisconsin studied an almond-shaped area deep within the brain—the amygdala—which functions to detect danger. The investigators found that autistic children have smaller amygdalas, and the smaller the size of the amygdala, the less willing the child is to make eye contact with others.

Some scientists believe that an autistic child's fear of socializing and making eye contact is what causes the amygdala to shrink. Other research, however, points to a more genetic component: normal siblings of autistic children also present a minor reluctance to make eye contact (not nearly to the degree of a fully autistic child). These siblings also tend to have slightly smaller amygdalas than normal children who have no family link to an autistic child.

Just as autistic children may have extraordinary intellectual and creative abilities, many young digital prodigies show similar achievements in select technological skills. Moreover, Digital Natives, after long periods of time on the Internet, display poor eye contact and a reluctance to interact socially. With the digital age evolving our brains, some experts argue that our society in general is becoming more autistic in the sense that people are spending less time interacting directly with others and more solitary time in front of their computers.

Functional MRI studies of young adults (aged eighteen to twenty-six years) who average fourteen hours a week playing video games have found that computer games depicting violent scenes activate the

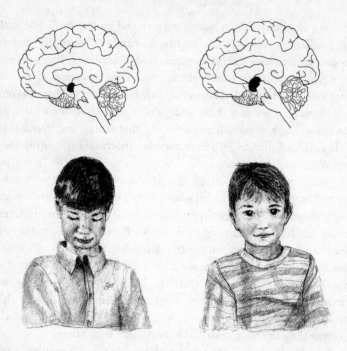

amygdala. It is perhaps no accident that many autistic individuals, with their small amygdalas and poor eye contact, are almost compulsively drawn to and mesmerized by television, videos, and computer games.

MYSTERY ONLINE ILLNESS

A new form of infection is spreading across the Web, but it's not a virus that shuts down computers. In this case, it is the mind of the individual user that gets infected. Victims experience a range of symptoms, from "brain fog" to chronic itching. The first outbreak was reported in 2001, when a two-year-old boy developed symptoms of "crawling, stinging, and biting sensations," along with skin lesions. His mother labeled it Morgellons disease, and after she established the Morgellons Research Foundation and its website (Morgellons.com), people throughout the world have reported these same "disease" symptoms.

Many sufferers believe that a new form of virus, worm, or parasite is afflicting their bodies. Other theories point to poisoned bottled water, alien forces, or toxic gases. In extreme cases, the dermatologist will diagnose delusional parasitosis (i.e., a false belief of being infected by parasites), which is not a new condition. What is new is the online spread of these symptoms.

Common complaints include clouded thinking and inability to work. The condition appears to be spreading at an alarming rate, but the symptoms and mode of contagion suggest that Morgellons disease is actually mass hysteria dispersed via the Internet.

In my report in the *New England Journal of Medicine*, I defined mass hysteria as an outbreak of illness that has a psychological rather than a physical cause. Typical outbreaks develop during school assemblies when children observe other children faint or develop some type of physical symptom. Then others in the group may start noticing symptoms in themselves, and soon the auditorium is evacuated so health officials can search for toxic causes. Although environmental contaminants may contribute to such outbreaks, most often anxiety and social contagion are the major contributing factors.

In any case of conversion hysteria, whether it affects an individual or a group, the mind of the victim converts an uncomfortable psychological conflict into a physical symptom. Some classic cases involve sudden onset of paralysis or blindness. Through the physical symptom, the patient temporarily avoids a psychological conflict by instead focusing attention on the body.

During my training in psychiatry at Boston's Massachusetts General Hospital, I treated a young man who was having escalating and heated arguments with his overly controlling father. As he was about to punch his father in the face, he suddenly went completely blind, and an ambulance rushed him to the ER, where we could find no physical cause of his blindness. Hypnosis and psychotherapy eventually helped him to see again and deal with his anger toward his father.

Functional neuroimaging studies of patients with hysteria have demonstrated areas of decreased brain activity in neural circuits that normally control the impaired physical function or sensation. Individuals with conversion hysteria who are experiencing muscular paralysis show decreased activity in the subcortical circuits (beneath the brain's outer rim) involved in motor control. Hysterical blindness causes deficits in the visual cortex; and persons with hysteria who experience numbness show deficits in the somatosensory cortex. At the same time that any hysterical physical symptoms and their corresponding neural circuits shut down, the emotional limbic brain regions show increased activity; this suggests that stress-related emotional circuits, not completely under the patient's conscious control, drive the symptoms.

Most outbreaks of mass hysteria spread when symptoms are observed directly in groups of children or adolescents, although media coverage of these epidemics has been found to spread psychosomatic symptoms and to kindle new outbreaks. In the past, newspapers, television, and radio were the major media of contagion. Although the Centers for Disease Control and Prevention have launched an inquiry to determine whether any disease clusters suggest an underlying organic cause, most experts agree that Morgellons disease likely reflects the impact of digital information transfer on the Internet as the latest mode of social contagion of this psychological illness. My prescription: turn off the computer and go outside.

CYBERSUICIDE

To some, the term "cybersuicide" may imply doing something incredibly stupid like accidentally pushing "reply all" to your entire office staff instead of just your buddy when sending out a joke about your boss. To others, it concerns the more than one hundred thousand websites that cover the topic of actual suicide methods and preparation. In 2004, cybersuicide was brought into the limelight when several Japanese adolescents who met over the Internet planned their suicides together through a website.

Some sites encourage suicidal thinking and behavior and may discourage people from seeking psychiatric help. Somewhere on the Web you can look up the best angle to point a gun in your mouth for the most lethal effect or find a list of toxic dosages of various drugs according to body weight. There are websites that list overseas pharmacies and ways to avoid legal problems when dealing with them. Others show graphic images of completed suicides or sample suicide notes. In suicide-related chat rooms, you are likely to read real suicide threats.

These chat rooms have facilitated suicide pacts through jumping and overdosing.

Young people are more likely to be risk takers and drug abusers, which may explain why a recent report found them at greater risk of being influenced by suicide websites than older adults. The challenge of cybersuicide is so new that policy makers have no strategy yet to address it. At the least, health care providers need to counsel patients about alternatives to surfing the Web at times of crisis. The NIMH (nimh.nih.gov/suicideprevention) and other national organizations (save.org; afsp.org; sprc.org) provide online information and resources to help families deal more effectively with suicide prevention.

I'M TOO TECHY FOR MY BRAIN

Technology not only affects groups of people large and small but also has a range of effects on individual behavior. Hyperactivity, inattention, depression, and multitasking mania are just a few of the behavioral consequences of the new techno-brain. With the average young person's brain exposed to a hefty eight hours of technology each day, the high-tech revolution likely has an impact on nearly every form of behavior. Although the science behind the way technology affects behavior and mental state is only in its infancy, initial observations indicate important links between extensive brain exposure to new technology and mental disorders.

Depression severe enough to require medical treatment afflicts an estimated 15 percent of the population at some point in life, and many people, particularly Digital Immigrants, note worsening of depression symptoms from too much exposure to technology. Previous studies have shown that social isolation clearly increases the risk for depression and worsens its symptoms. Despite the availability of social networks, email, and instant messaging, these electronic communication modes lack the emotional warmth of direct human contact and often worsen a person's feelings of isolation.

The high-tech revolution also has contributed to many forms of anxiety, ranging from chronic generalized anxiety disorders to panic attacks that can be disabling. Baby boomers and seniors complain about computer anxiety and fear the dangers of the Internet, not only for themselves but also for their children and grandchildren. Patients with obsessive-compulsive disorders often find that when they get involved with digital technology, whether it's email, shopping, or video gaming, they cannot control their impulses to continue and become addicted (see Chapter 3). The brain circuitry and maladaptive behaviors that control addictions and compulsions show considerable overlap.

Patients suffering from these and other behavior disorders can also get help from new technology. Mental health education, blogs, and Internet support groups are available for people with depression, obsessive-compulsive disorders, panic attacks, and nearly any type of psychiatric condition (see Appendix 3). They simply have to want to find the help.

Fundamentally, the new high-tech world is influencing how young people develop their sense of self and worth in this world. The ability to immerse oneself in a fantasy-universe game, or the empowerment of being just a few keystrokes away from any acquaintance throughout the globe, is shaping—both for good and for ill—each young person's identity and self-esteem: essential elements that dictate people's actions and define their humanity. For older Immigrants, new technology helps them remain effective at work, stay in control of their lives, and keep their hands on the pulse of today's culture.

Five

HIGH-TECH CULTURE:
Social, Political, and Economic Impact

> *For a list of all the ways technology has failed to improve the quality of life, please press three.*
>
> Alice Kahn, *author and journalist*

As digital technology permeates almost every aspect of our lives, it transforms our social, economic, and political worlds. Most of us are unaware of how our brain's neural circuitry is evolving and responding to this transformation, because many of the changes in our everyday experiences are very subtle. For example, in just the past decade, traditional low-tech financial transactions, such as writing personal checks, have declined by nearly 50 percent, while the number of electronic payments has tripled. Our brains adapt to such cultural changes, and we respond emotionally as well. These reactions can range from the negative—fear of cyber crime, loss of privacy, and technology fatigue—to the positive: increased efficiency, enjoyment of vast entertainment options, and a greater sense of control.

MULTIPLE CHOICE

A 2005 Pew Internet survey found that 45 percent of Internet users— approximately sixty million Americans—claimed that the Web helped them make major life decisions or negotiate challenging life events: 54 percent used the Internet to help them cope with physical illnesses, 50 percent noted that it assisted them in pursuing career training, 45 percent found help with financial or investment decisions, and 43 percent got assistance when looking for a home or apartment. Whether it's a major life decision or an impulsive entertainment selection, today's techno-culture is offering us more choices than ever before, and our brains are adapting to this abundance of choices.

When Digital Immigrants started out in the workforce, they hung out at the water cooler to gab about the previous night's television show on one of the three network TV stations. Mega-hits and celebrities still capture public attention, but the Internet now lets us pursue whatever personal curiosity intrigues us. And, we don't need to hang out at the water cooler to discuss those interests. Broadband connections and social networks let us chat about our pastimes with like-minded individuals across the globe whenever we feel like it.

This drift from the mega-hit mentality to a personal interest focus is altering the strategies of marketers and ad agencies, both traditional and Web based. In his book *The Long Tail, Wired Magazine* editor Chris Anderson describes how companies selling online products, such as books or DVDs, don't need to worry about shelf space because they can advertise huge inventories of slow-selling products and simply fill orders as they come in. This long tail of customer options looks to be more profitable over time than pushing fewer but more popular titles or products. Consumers are given a lot to choose from—they can move beyond the bestseller stalls and use a search engine to troll for anything they want, no matter how obscure.

Anderson notes that approximately one-third of Amazon sales consist of books that you would never find stocked on a major bookseller's shelf. It's not yet clear how much this infrequently ordered, less popular inventory is driving Web businesses. For example, 3 percent of Amazon products generate 75 percent of its profits. Anyone who has searched the Web knows of the extraordinary choice it provides. It's not just products that we can access instantly, but any form of information we may want, whether the address of a restaurant or an obscure fact needed to settle a bet.

Celebrities, mega-hits, and strong brands have always been big sellers. In fact, our brains are hard-wired to seek out established brands. Dr. Christine Born and colleagues at Ludwig Maximilians University in Munich used functional MRI scanning to show the brain's response to strong and weak brands in research volunteers in their late twenties. Areas in the frontal lobe of the brain that control positive emotions—the

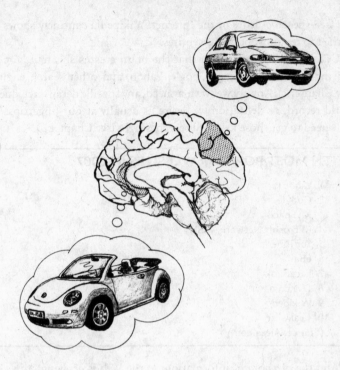

insula and the anterior cingulate—lit up in response to a strong car brand, Volkswagen, but not to a lesser-known brand, Seat. The weak brand provoked a different brain region associated with negative emotion, known as the precuneus.

INFINITE INFORMATION

In this era of unlimited choice, the door-to-door encyclopedia salesman has vanished—we no longer need expansive sets of multivolume reference books when sites like Wikipedia (a Web-based, free-content encyclopedia written collaboratively by volunteers), Google, and Ask.com are so accessible. Your computer-savvy son or granddaughter is not going to ask for an actual bound almanac or thesaurus.

Author Andrew Keen describes the negative cultural impact of such editor-free sites, but his concern has little impact on the popularity of the free-speech culture of the Internet. Wikipedia routinely shows up in the top ten most popular websites.

UCLA studies have found that the brain creates shortcuts for acquiring information from Google, Yahoo, and other search engines (see Chapter 1). Any answer we may need, any specific detail, fact, quote, world record, or definition we desire, is usually at our fingertips. We only need to get those keyword searches right (see Chapter 8).

TEN MOST POPULAR WEBSITES IN 2007

1 Yahoo!
2 Google
3 MySpace
4 Microsoft Network (MSN)
5 YouTube
6 eBay
7 Facebook
8 Live.com
9 Wikipedia
10 Craigslist
(Source: Alexa.com)

And the plethora of information on the Web is growing daily. The Internet monitoring company Netcraft counted 122,000,635 websites at the beginning of June 2007, which represented a gain of nearly four million sites since the previous month.

The Internet empowers us to track—from moment to moment—cultural interests, book sales, public opinion, political contributions, and even local crime rates. We can shop, snoop, study, or search until our fingers get stiff and our eyes become blurred. Then we can search out how many other people have stiff fingers and blurred eyes.

Systematic reviews of people's search habits have revealed some consistent patterns. For example, a recent analysis of 650,000 AOL customer searches indicated that Web searchers seek entertainment first and shopping second. The Pew Internet & American Life Project reported that 80 percent of American Internet users search for health-related information.

In a 2006 *Wall Street Journal* article, Lee Gomes contrasted some intriguing samples of random Web searches: the keywords "Britney Spears" triggered more than enough results to outrank the keyword "God." In January of 2008, my own Google searches indicated that the term "God" was holding its own at 551 million results, easily outranking the 66 million for "Britney Spears," but not nearly approaching most people's favorite search term, "free," which logged in an impressive 4.9 billion results. One can get a snapshot of mass interests at any point in time just by querying the top Internet search terms by date: shopping, entertainment, hobbies, and celebrities routinely make it to the top ten (see Table).

The top ten search terms of 2007 for Google, Yahoo, and Ask clearly

TOP 10 SEARCH TERMS FOR 2007

Google.com	Yahoo.com	Ask.com
1. iPhone	1. Britney Spears	1. MySpace
2. Webkinz	2. WWE	2. Dictionary
3. TMZ	3. Paris Hilton	3. Google
4. Transformers	4. Naruto	4. Themes
5. YouTube	5. Beyonce	5. Area Codes
6. Club Penguin	6. Lindsay Lohan	6. Cars
7. MySpace	7. Rune Scape	7. Weather
8. Heroes	8. Fantasy Football	8. Games
9. Facebook	9. Fergie	9. Song Lyrics
10. Anna Nicole Smith	10. Jessica Alba	10. Movies

differed. There's only a single overlap: "MySpace" appears on both Google and Ask. Although the methods and criteria for ranking searches differed among the three sites, this comparison also suggests that different types of people, with differently wired brains, use each of the three engines. Also, the style and presentation of a particular site will affect a person's choice. However, nearly all Web searchers enjoy the fact that they can not only find a wealth of information but usually get it for no charge. It's no surprise that one of the most frequent database search words is "free." This pervasive attitude of "find what you need without having to pay for it" partly explains why many Internet-based businesses have filed for bankruptcy—they were selling information or links to products that Web browsers could often find elsewhere on the Internet, free of charge.

THE ELECTRONIC MARKETPLACE

Initially, online shoppers tend to stick to a few purchasing sites. They begin with items that meet their basic entertainment needs, such as books or music. With greater Internet experience, they begin to venture out to new sites—at least 75 percent of online shoppers are willing to expand their purchasing repertoire and sample new sites. A study from the network information provider comScore Networks reported that in 2005, the amount of money spent shopping online totaled $83 billion, indicating a 24 percent increase from the previous year.

Internet users are not only shopping online but selling as well. One out of every six Americans sells something online; according to the Pew Internet & American Life Project, on a typical day, 2 percent of Internet users sell online. Those most likely to sell are daily Internet users between thirty and forty years old, as well as people who have longer Internet experience. The survey found that nearly one out of four people who have used the Internet more than six years report selling on line. This phenomenon of consumers becoming vendors involves brain adaptation, wherein new neural circuitry is being laid down.

The ability to search efficiently for a range of products and compare prices quickly makes the online marketplace particularly attractive. Although shopping for small items is the norm, the Web also provides opportunities to shop for big-ticket items, including cars and real estate. According to the National Association of Realtors, approximately 80 percent of buyers use the Internet to assist them in finding a home. Websites like Zillow.com post the estimated values of millions of homes, and people have been reported to make home purchases based solely on virtual tours, never having visited the property before the sale closes. Automotive websites have developed search engines that allow shoppers to compare prices and models according to horsepower, cargo size, fuel economy, and many other features. Some of these sites allow you to configure your car with exactly the features you want, and then find a dealer in your area who has that exact car in stock.

Whether we're bargain hunters or big spenders, a specific area of the brain, the nucleus accumbens, helps us make decisions while shopping. Dr. Brian Knutson and colleagues at Stanford University used functional MRI scans to predict, via brain activity patterns, whether subjects would or would not buy something, even before that person had made a conscious decision about buying the item. When subjects viewed an online image of something they really wanted to purchase, the nucleus accumbens was activated.

This brain region contains numerous dopamine receptors that are stimulated during pleasurable experiences or when pleasure is anticipated. However, if an item is overpriced or damaged, a completely different part of the brain, the insula, fires up. This insular region regulates unpleasant experiences, like smelling a foul odor or seeing something gross. It is no surprise that overspending or viewing a large credit card bill activates the same neural circuitry that controls pain and discomfort.

Advanced digital technologies have created a more efficient marketplace, and consumers can focus their purchasing power on only the products they truly want. Gone are the days when we would have to buy an entire roll of twenty-four or thirty-six photo prints, when all we really wanted were those five great snapshots that captured the moment just right. Now we can save or delete as many digital photos as we like for no extra charge. Music technology lets us download a favorite song without purchasing the entire CD, and advertisers can focus their marketing campaigns on target audiences through MySpace, YouTube, and other popular sites, since many charge a premium for their ads only when consumers choose to click on them.

WEBONOMICS

Economic experts point to technology's positive impact on productivity and standard of living. The efficiency of the Web reduces the costs of transactions needed for producing and distributing many products and services. It's easier for consumers to comparison shop for an item among many vendors; this results in increased competition, which allows for greater savings, choice, and shopping convenience. You can find several prices for the same hotel room on the same dates at a particular resort, depending on whether you visit Travelocity.com, Expedia.com, Hotwire.com, or one of several other travel websites. Routine transactions, such as making a mortgage payment or transmitting financial information, require less time and expense using Web-based technology.

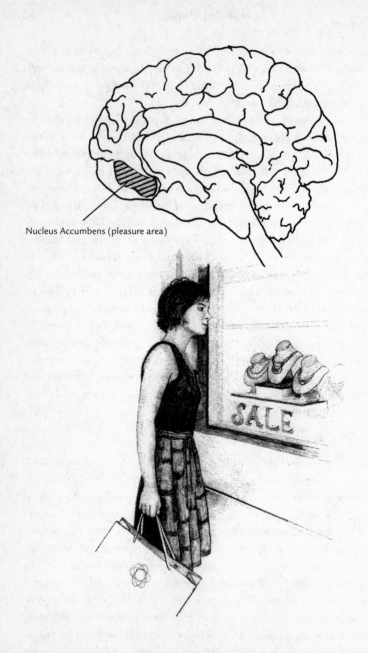

Nucleus Accumbens (pleasure area)

The fact is that we are becoming a cashless society. Almost everything can be paid for electronically or by credit card or debit card. *Newsweek* reported that in 1990, Americans rarely if ever paid for groceries by credit card. In June 2007, according to the Food Marketing Institute, credit cards were used to pay for approximately 65 percent of all food purchases.

Mobile banking has clearly come of age. Basic banking activities can now be accomplished from hand-held devices. Financial institutions also offer online tools to help customers manage their money and investments. Users can review their spending habits, create budgets, and receive email alerts when they are about to exceed their limits or haven't paid bills on time. Eighty-three percent of Social Security beneficiaries have their monthly payments automatically deposited into their bank accounts. These direct deposits have reduced the number of stolen checks from seniors' mailboxes; however, this method is not foolproof, because computer hackers and identity thieves can get their hands on personal banking and Social Security information. Many banking websites have protections in place to thwart cyber criminals (see Chapter 8).

We are seeing technology's impact not just in the new e-commerce but also in health care. Any pharmacist, physician, or other health care professional can go online and review a patient's health and medication history dating back several years. Fed up with the inefficiencies of group practice, family physicians are harnessing technology to make their solo practices more profitable and manageable. Using electronic medical records and streamlined billing programs, and offering email consultations, doctors are now able to lower their overhead and spend more time with their patients as well as their own families. Bidshift Inc., a San Diego software company, has developed a program to help hospital administrators cope with escalating overtime costs. The computer program lets nurses bid for preferred shifts so that personnel costs better match personnel preferences.

Search engines and other new technologies have made the Internet big business. The Interactive Advertising Bureau reported that online advertising totaled $16.8 billion in 2006, a 34 percent jump from the previous year. Brokerage firms have begun using computer programs to automate complex stock trading strategies without the need for input from actual brokers.

The Web can also be used to help you get a raise. Sites like PayScale .com and Salary.com allow users to check comparable salary ranges that other people receive in the same job position. Armed with such information, employees are requesting pay increases from their bosses (assuming, of course, that they are underpaid).

Recent economic analyses show that the more effectively we use new information technology, the greater the economic gains. The challenge is to create incentives for greater use, particularly in health care and transportation. For example, a national health information network could create tremendous financial savings by sharing information about patients' diagnoses and treatments. Many people, however, remain concerned about the privacy issues involved in having their medical records become public on the Internet.

When we make decisions about our finances, whether it's online or off line, our brains draw on the same neural circuitry used whenever we choose or avoid any kind of risk in life. Unfortunately, our brains are not always rational when it comes to financial choices, and economists have described two major errors that people tend to make in their financial decisions. Some become too cautious in order to avoid risk, while others go overboard, seeking high profits and taking too much risk. Functional MRI scans have shown that risk-seeking financial errors stimulate the nucleus accumbens, the dopamine-rich area that drives our shopping instinct, while decisions based on risk avoidance activate the anterior insula, the region that fires up if we view a very large credit card bill.

Other brain regions also control our economic decisions. Feelings about future economic uncertainty plague many investors in the stock market, real estate, and other areas. Neuroscientists have found that such uncertainty triggers neural circuits in the amygdala, which integrates our thoughts and feelings. At the same time, the orbitofrontal cortex controls vigilance, a mental process necessary for maintaining a watchful eye on market volatility. A typical outcome when fear of the economic unknown activates these circuits is for the investor to prefer the familiar. Though the strategy may calm down our brains' neural pathways, it may cause us to hold on to unwise investments rather than diversify our stock holdings toward a safer or possibly more profitable portfolio.

SOCIAL NETWORKING AND ENTERTAINMENT

In part because of Webonomics, our society has become electronically connected. With the ease of text messaging and email on portable hand-held devices, along with almost constant exposure to some form of electronic media, it's possible that very soon the only time people will be off line is while they sleep.

The Internet is becoming a major source of news and current events, and traditional print media outlets are feeling the crunch. The Newspaper Association of America has been noting an increase in newspaper website use in recent years, while daily hard-copy newspaper circulation is declining nationwide. In early 2007, *Time* magazine and other popular print weeklies made major cutbacks in their hard-copy publication budget and boosted their online presence.

The music, television, and film industries have long been undergoing technological transformation. Rather than go out to the movies or the theatre, many people prefer the convenience of pay-per-view movies or digital recordings of their favorite shows—watched in the comfort of their living room on their sixty-inch flat screens, or perhaps snuggled up on an airplane with a laptop or hand-held device.

With fewer customers visiting traditional media outlets like network TV and movie theatres, entertainment companies are reaching out to cyber-social networks to capture their audiences. YouTube has licensing deals with thousands of partners, ranging from cable channels like Sundance to small independent producers and even teenagers with a video camera. The potential for mass distribution on the Internet is particularly appealing to start-up content providers. In the fall of 2007, seasoned film and TV producers launched their first regular television series on MySpace, along with a website to promote the show. Traditional TV watchers will still be able to view the series on their sets, but only when the show goes to reruns.

Advertising revenue from Google, YouTube, MySpace, Yahoo, Facebook, and other online companies has grown steadily along with the number of viewers. In the United States, this revenue increased to nearly $17 billion in 2006—a 34 percent rise from the previous year. Just a decade ago it would have been considered absurd for a major broadcast company to contemplate the purchase of a website that posted personal ads, homemade video clips, and private ramblings. But in 2006, Rupert Murdoch's News Corporation, owner of FOX Networks, bought MySpace for $580 million, and Google purchased YouTube for $1.65 billion. To reach the public, corporate America has gone to the users—and they're getting their money's worth.

MySpace, Facebook, and similar sites are more than just social party sites that teenagers and college students get hooked on. They have become influential marketing forces. Large corporations, entertainment suppliers, and politicians are harnessing this power source. These online communities have become destinations for niche interests, where users can blog and share their musings on anything from their favorite sci-fi movies to their own particular political point of view. Health advocacy groups are now testing online communities as a way to disseminate information on disease awareness and prevention.

Driving down to the video store to rent a tape or DVD will soon become an anachronism of the twentieth century. With a few keystrokes we can select our choice of home entertainment in seconds. The mobile content-streaming service MobiTV now allows consumers to watch high-quality videos on cell phones, iPods, and other small devices. It's becoming a digital entertainment feeding frenzy. All this access, and the vast selection, is causing some entertainment seeker's brains to develop a need for instant gratification, a loss of patience, and an actual change in neural circuitry—a sort of entertainment ADHD. Some people can barely finish watching a half-hour program for fear they are missing too many other options.

Some YouTube video clips are viewed more than a hundred million times, and this worldwide free-expression medium has given audiences the upper hand. If you missed a media event on TV last night, don't fret—you may be able to catch all or part of it posted on YouTube.

Such rapid advances in technology are undermining the movie industry's traditional economic model, which depends on large opening weekends and DVD sales. As the online audience demands greater control of the entertainment content, the suppliers will be forced to adapt and accommodate.

WOMEN VS. MEN ONLINE

Evolution has programmed men and women to behave differently, even in the way they use and respond to technology, and a recent survey from the Pew Internet & American Life Project highlights such differences. Women are more likely to email friends and family to share concerns, forward jokes, or plan events. They value the way the Internet enriches their relationships. Women tend to seek diet and health information on the Web and to worry more about criminal Internet threats. Women are also more likely to download online map directions than are men. By contrast, men frequently log on to the Web for news and financial updates, sports scores, and video games. The Pew Study also showed that men tend to be more tech-savvy, more confident in using search engines, and more likely to try new gadgets and software.

Evolution-driven adaptations in brain structure and function might explain some of these differences between men and women in technology habits. The female brain is hard-wired to take in the big picture, recognize subtle and not so subtle social cues, and experience more empathy. Women also tend to be better than men at reading people's moods. By contrast, men's brains are hard-wired to focus more on small details and to grasp visual and spatial concepts more readily. The male brain also tends to operate with greater emotional detachment.

Such gender differences in brain function remain apparent even when brain disorders profoundly affect an individual's ability to connect socially. For example, Asperger's syndrome and autism, conditions that impair human contact skills, are more common in males than females. Dr. Simon Baron-Cohen and colleagues at Cambridge University have shown that higher levels of the male hormone testosterone early in development are associated with lower ability to make eye contact at one year of age. Higher prenatal testosterone levels also correlated with lower vocabularies at eighteen months and twenty-four months of age. Baron-Cohen concluded that in a sense, the autistic brain is a more male brain.

Dr. Richard Haier and colleagues at University of California in Irvine studied gender differences in the brain according to IQ scores. They found that a larger volume of gray matter (the neuronal cell bodies), distributed throughout the brains of men, was associated with higher IQ, whereas for women, higher IQ came from the brain's white matter (the axons or wires that connect the cell bodies), concentrated in the frontal lobe. This centralized frontal intelligence processing in women is consistent with other studies showing that women are more sensitive to frontal brain trauma than men, and it explains a woman's advantage in taking in the big picture of complex social situations.

At the Internet's inception, men initially dominated online activity, but the Pew Internet & American Life Project found that in the past two decades, women have increased their use and leveled the playing field. Despite such trends, gender differences in the patterns of technology use persist, and the differences in brain function and structure between the sexes suggest variations in brain vulnerabilities. Excessive exposure to digital technology may make the male brain more likely to exhibit autistic-like behaviors—poorer eye contact and less ability to make empathic connections.

FRACTURED FAMILIES

Investigators at the University of Minnesota found that traditional family meals have a positive impact on adolescent behavior. In a 2006 survey of nearly a hundred thousand teenagers across twenty-five states, a higher frequency of family dinners was associated with more positive values and a greater commitment to learning. Adolescents from homes having fewer family dinners were more likely to exhibit high-risk behaviors, including substance abuse, sexual activity, suicide attempts, violence, and academic problems. In today's fast-paced, technologically driven world, some people consider the traditional family dinner to be an insignificant, old-fashioned ritual. Actually, it not only strengthens our neural circuitry for human contact (the brain's insula and frontal lobe) but also helps ease the stress we experience in our daily lives, protecting the medial temporal regions that control emotion and memory.

Many of us remember when dinnertime regularly brought the nuclear family together at the end of the day—everyone having finished their work, homework, play, and sport activities. Parents and children talked, shared their experiences, and kept up with one another's lives.

Now, dinnertime tends to be a much more harried affair. What with emailing, video chatting, and TVs blaring, little time is set aside for family discussion and reflection on the day's events. Conversations at meals sometimes resemble instant messages, whereby family members pop in with comments that have no linear theme. In fact, if there *is* time to have a family dinner, some family members tend to eat and run back to their own computers, video games, cell phones, or other digital activity.

Tricia had the afternoon off that Friday, so she decided to cook her family's favorite dinner—turkey with all the fixings. When her fifteen-year-old daughter, Jenny, called to see if her friend Kali could sleep over after school, she said sure—they certainly would have enough food. Tricia searched online for a recipe she had seen in a magazine the previous Thanksgiving, and printed it out. She would have emailed her shopping list to the market for delivery, but she wanted to get the bird in the oven ASAP, so she went to the store instead.

When she finally got everything baking in her new computerized convection oven, Tricia set the bedroom TiVo to record a movie she wanted to watch later that night with her husband, Greg. Tricia's twelve-year-old son, Max, came home from school and immediately got deep into a video game in the den. She stood in the doorway and told him she was making his favorite dinner, but she didn't think he'd heard her.

Jenny and Kali got home soon after that and made a beeline for Jenny's room. Tricia asked if they had any homework, and somehow the girls didn't seem to hear her either. Was her entire family becoming hard of hearing? Jenny called over her shoulder, "Okay if I use your laptop, Mom?" Tricia asked, "What's wrong with your computer?" "Kali's going to use it." Tricia figured they did have homework after all and were going straight at it, so she replied, "Okay."

With the kids settled and dinner cooking nicely in the oven, Tricia went to take a relaxing bath before Greg got home. Then she checked the turkey, and it was perfect. She was setting the table when Greg came in from work and kissed her cheek. He said he had to sync his PDA to the desktop computer upstairs. He *promised* not to even open his email server. She sliced the turkey, put out all the sides, and yelled out for all to hear (hopefully): "Dinner is served! Come on, everybody!" Greg called down, "One sec, hon, I just have to do one thing." Tricia went into the den and saw Max still playing the video game. "It's dinnertime, honey.

(continued)

(continued)

Come on." "In a minute, Mom," he replied, and kept punching buttons frantically. "I'm really doing great."

She walked toward Jenny's room and heard the girls laughing. She smiled. It reminded her of when she was a teenager and had friends sleep over. They'd talk and laugh and gossip for half the night. But when Tricia went in, the girls weren't laughing with each other at all. Sure, they were sitting side by side, but each was on a separate computer, instant messaging and laughing with different friends. They really didn't even need to be together. "It's dinnertime, you two. Let's go." Grudgingly, Jenny and Kali followed her to the kitchen.

Tricia stopped by the den and shut off the television: "Right now, young man." "Okay!" Max harrumphed.

The only one not at the table was her husband, who of course had not resisted checking his email. "We're starting without you, Greg," she called up to him. "Be right there," he yelled down.

After waiting another five minutes and still no Greg, Tricia served the kids and herself, and asked them each about their day. Max shoveled his food in as if there were no tomorrow, and before Greg even came downstairs, Max asked to be excused. He jumped up and ran back to the den to restart his game.

"Greg! Come on, we're almost finished down here!" Tricia yelled upstairs, but this time he, too, didn't hear her. The girls asked to be excused and cleared their plates. They had to get back to their separate but together IM'ing. Greg called down from upstairs, "I have one more *really* important email, honey, and then I'll be down. I swear." Tricia had to laugh—she'd heard *that* a million times. So she finished her dinner, cleared the table, and went into the bedroom to watch the movie she'd TiVo'd earlier.

Greg came downstairs. "Okay! Where is everybody? I'm starved!" He could hear Tricia laughing at the movie in the bedroom. He spoke louder: "Honey? Where's the dinner?" Tricia called back, "It's all in Tupperware in the fridge. Make yourself a sandwich. Oh, and while you're at it, could you bring me a glass of wine?"

Tricia's situation is much like that of millions of other baby boomers whose family lives have been fractured by the ever-increasing presence of new technology. Tricia's effort to bring her family together over their favorite dinner was ineffective when she was up against the overwhelming draw of new technology. In the end, she coped with her disappointment by using her own favorite form of digital escape—TiVo.

Although the traditional dinner can be an important part of family life, whenever surly teenagers, sulking kids, and overworked parents get together at the dining table, conflicts can emerge and tensions may arise. However, family dinners still provide a good setting for children and adolescents to learn basic social skills in both conversation and dining etiquette. The opportunities for developing the brain's neural networks that control these skills are being lost as families become more fractured.

LOVE AT FIRST SITE

Romance is no longer as simple as spotting someone across the room, exchanging flirtatious banter, dating, and eventually moving in together or tying the knot. Now it often starts when you log on to a singles' website and search for that perfect someone—or at least someone who meets your requirements. Then you may email and IM each other until you finally hook up on MySpace, share your YouTube videos—and then, if your other twenty-five online friends agree, you might eventually meet in person and hope for the best.

Our capacity to network socially online has definitely jump-started the matchmaking business. A 2006 Pew Internet study found that nearly 40 percent of Americans who are single and looking for a relationship have used online dating services, such as Match.com or Yahoo Personals. The study found that a majority of the online dating experiences were positive. Nearly 20 percent of the subjects reported entering a long-term relationship or getting married through online dating.

The electronic approach to finding a mate has several advantages. Daters can rapidly review profiles and focus on someone who shares their interests and values. They are also able to meet people they might not otherwise meet because their social or business circles don't intersect. However, there are risks. In addition to concerns about privacy arising from the posting of personal information on the Web, some people who use these matchmaking services are dishonest about their profiles, get hooked on Internet flirting, or take advantage of others who are sincere about their own dating interests. Also, without face-to-face initial meetings, people are denied the use of their natural instincts to guide them on whether there is any personal chemistry that can drive a romance. A digital photo or video cannot convey the nuances of appearance, body language, and facial expression that give us emotional clues about another person. That all has to wait until a face-to-face meeting takes place, often following considerable time spent online.

The traditional love note—a familiar medium for expressing romantic feelings—has also gone electronic. According to the Greeting Card Association, 14 million e-Valentines were sent in 2007, and that number appears to be growing each year. We can now express our true feelings of love and do it on the run, using text messages, emails, and e-attachments of choice. "OMG . . . ILU :)" says it all.

Digital Natives, often craving instant gratification, seem to have little trouble with this new way of saying "I love you." Electronic love notes can be just as exhilarating as the hand-written variety, and they are instantaneous. No need to wait for the postman's delivery; just turn on your hand-held device and send and retrieve your email or text messages. Unfortunately, such love expressions can become public online without delay. Electronic lovers have been known to share their personal notes with friends, as well as solicit opinions about the sincerity and quality of the notes. Before long, tender, private expressions of affection can get posted on Facebook or MySpace. With this new form of love letter, the old rituals of treasuring handwritten notes is lost. Also, if your cell phone breaks or you erase your hard drive, these expressions of your sweetheart's deepest feelings can disappear in an instant.

Recent brain scan studies show that love dramatically changes the brain in a way that resembles drug addiction or obsessive-compulsive disorder. A picture of one's lover will excite the brain's dopamine system, which controls pleasure and addiction. But even months or years after a relationship ends, neuroscientists can observe that same dopamine system firing. This is not surprising, since many people have intense feelings that linger long after a romance formally ends. However, failed relationships also trigger brain areas that mediate risk taking, anger expression, obsessive-compulsive symptoms, and physical pain.

Dr. Helen Fisher of Rutgers University has pinpointed the brain's romance neural pathways. She argues that love is not just an emotional state but a mental experience that resembles addiction: it drives the brain to seek rewards through its dopamine neural circuits. Also, as with any habitual or obsessive behavior, a series of complex actions orchestrated by the prefrontal cortex drives people to great lengths to feed their habit—often compelling them to do things they would not normally do.

The prevalence of portable technology also threatens traditionally intimate environments. Many tech-savvy couples are sharing their beds with a third party—a spouse's laptop, BlackBerry, or iPhone. Some are shocked by the intrusion—they see the bedroom as a sacred chamber where they can escape from the outside world. Other couples think nothing of it—they happily write their blogs well into the evening hours and claim that laptopping in bed is no different from watching television or reading. With at least 30 percent of Americans now owning laptops, we can expect that the clicking and whirring of hard drives will become as familiar a bedtime sound as a spouse's snoring.

TECHNOLOGY AND PRIVACY

During the summer of 2006, a momentary computer glitch illustrated the Internet's threat to personal privacy on a grand scale. AOL inadvertently posted nineteen million Internet search queries from more than six hundred thousand customers.

Whenever we search for information on the Web, we run the risk of revealing our identities to predators, identity thieves, or government watchdogs. Each time we buy something online or sign up for a free service, that website creates bits of information, or cookies, that uniquely identify our computer and jeopardize our privacy. Security threats are even greater when we use wireless services, and wireless Internet access is expanding. Entire cities are going Wi-Fi for free. This is creating another growing threat from fake Wi-Fi hotspots created by cyber criminals who wish to steal financial and other sensitive information from personal computers.

Search engines like Yahoo and Google store data on a user's search words, as well as the specific computer and browser that is used for the query. The more we search, the more we reveal our particular tastes to online marketers. By observing a shopper's digital trail, these Web merchants can target their strategies to match products with customers. For instance, if you click on one or two mortgage broker sites, loan offers and solicitations from several other mortgage sites may arrive by email in the next few hours.

Retailers and service providers are not the only concern stressing out the brains of today's computer users. Any individual can run a search and find detailed information on other people—where a person works, email or home address, religious affiliation, and even the amount of donations to a political party. Property tax assessments, motor vehicle records, and voter registration files can be found online as well. Posted court records may contain social security numbers and financial account numbers, posing risks of identity or financial theft.

Computerized health care records create another potential privacy hazard. When high-profile celebrities are admitted to the hospital, computer hackers routinely attempt to pry into their records. Despite the convenience and efficiency of online records, only an estimated one in four practicing doctors in the United States uses electronic health care files, often because of privacy and other concerns. Many doctors avoid emailing with their patients because it's impossible to always protect the privacy of these communications, which would violate the HIPAA (Health Insurance Portability and Accountability Act) regulations. Emails also take up time, and few health insurance carriers reimburse doctors for that time. Currently there are Web-messaging tools available to protect patient information, and some health professionals are beginning to use them. The federal government's National Technology Health Information plan has set a goal for all Americans to have electronic health records—but not until 2014.

It's not just criminals snooping into our lives that we have to worry about, but employers as well. Bosses are getting wise to employees' personal use of computers and phones during company hours. Many businesses are now using programs like Xora and SurfControl to monitor their employees' electronic behavior, tracking anything from email gossip to eBay trading. Some companies, particularly those with several service technicians, have begun using GPS technology to track workers' whereabouts while they're talking on company cell phones.

Many students haven't a clue that those provocative postings and photos on their MySpace pages can be visible to strangers, potential employers, and college admissions officers. Ethicist Randy Cohen notes that reading a student's blog is legal, but that it is unfair to use this information against a prospective enrollee or job applicant. Whether on MySpace, Facebook, or any other social networking site, anyone who registers can view personal pages with minimal restrictions.

In 2006, the career of *Seinfeld* star Michael Richards (Kramer) took a nosedive after his racial slurs during an improvisation performance were immortalized by a video camera phone and posted on the Internet. Miss Nevada USA lost her crown after photos of her indiscreet partying were posted on the Web.

Digital cameras are now standard features on most cell phones. In 2008, an estimated 80 percent of all cell phones sold in the United States will have such cameras. These portable electronic gadgets come in handy for remembering new acquaintances, documenting a fender bender, or perhaps catching a snapshot of a crime in progress. They also provide angry audience members, surreptitious students, and jealous lovers an opportunity to capture compromising moments that can then be disseminated instantaneously on the Web. Seeing a still or moving image can have a much greater impact than reading something in writing. Anything we say or do can now become public record, and as yet there are very few camera-free zones for those who wish to let their hair down. Some gyms have banned cameras and cell phones from their locker rooms, but how can that really be monitored without a camera?

CYBER CRIME

Owing to the growth of the Internet and new tools to carry out cyber crime, criminals have numerous opportunities to conduct online attacks, and the frequency of Internet crime is on a steep rise. The FBI ranks cyber crime among its top three priorities, just behind terrorism and espionage.

Political terrorists have found a way to use the Web's anonymity to expand their networks and temporarily avoid detection by law enforcement agencies. In 2006, there were an estimated five thousand Internet sites where terrorists were recruiting and radicalizing a post–9/11 group of cells. This new generation of terrorists is using the Internet to create a global network that can facilitate communications in an instant.

Political extremists also use the Internet and video technology to spread worldwide fear and shock. Websites posting gruesome images of beheadings and torture are sending messages of fear and intimidation, which have a profound impact on the emotional centers of the brains of all those who view them. Dr. Jack Nitschke and colleagues at the University of Wisconsin have shown not only that viewing gruesome images activates a specific network of cognitive and emotional brain regions, including the prefrontal cortex, anterior cingulate, insula, and the amygdala, but that merely anticipating these types of images triggers the same neural circuitry.

Some computer hackers deceive their victims by using fake websites that mimic known brands. The use of these fraudulent sites, known as phishing, dupes thousands of people each day into buying fake or nonexistent goods and inputting their credit card information.

Cyber police are now using digital technology of their own to pursue and prosecute Web-based offenders. Worldwide cooperation is often crucial for catching computer hackers who set Internet viruses and worms in order to disable computer networks and steal credit card and social security numbers. According to the research firm Computer Economics in Irvine, California, malicious computer viruses caused an estimated $14 billion in damages in 2005. That year, the FBI worked closely with law enforcement agencies in other countries to arrest Turkish and Moroccan hackers who created the Zotob virus, which snatched credit card numbers and infested computers at CNN, the *New York Times*, and an estimated 100,000 other companies. To apprehend the criminals, the cyber police tracked Internet protocol addresses (unique sets of numbers) from emails sent by the hackers and used evidence retrieved from reformatted disk drives.

In China, citizens use text messaging to report a crime when phoning might endanger their safety. The Boston Police Department has established a "Text a Tip" program to help police catch criminals. Text messages also have been used during Amber Alerts to broadcast child abductions. Some police departments are considering general emergency warning systems that use text messaging as well.

As technology continues to become more complex, cyber criminals will probably continue to come up with new ways to scam innocent people. Fortunately, fresh innovations are consistently being created to help catch these offenders.

I'D RATHER BE BLOGGING

New voices are being heard online every day through Internet blogs (Web log = blog: a user-generated website where entries are made in journal style and displayed in reverse chronological order [Wikipedia]). Anyone can create a blog to voice his or her views to the world. In 2007, an estimated seventy million blogs existed—about double the number from the previous year.

Although blogs are often associated with politics or entertainment themes, it turns out that 54 percent of bloggers are simply individuals who use the format as a personal journal. Bloggers generally seek creative expression and wish to share their personal experiences with others. The Pew Internet & American Life Project found that most bloggers are under age thirty, and they blog as a hobby, not a profession.

Bloggers can create and edit posts and upload videos and photos from their hand-held devices. With the opportunity to constantly update their sites, some bloggers share nearly every private moment and often reveal what many consider to be too much information. Such "live bloggers" sometimes stop living in the moment and instead spend their time memorializing every little detail of their daily lives.

Marketers and publicists recognize the power of the blog. They are developing their own blogs to add value to their products and services while using other blogs to promote and discuss them. Yahoo and Google pay experts to maintain a presence in the blogosphere. Music promoters use blogs, as well as social networking sites, to push their latest releases toward the top ten lists, while Hollywood studios enlist bloggers to help promote movie and Oscar campaigns. Because advertisers are buying bloggers' voices, blogging-for-dollars raises ethical questions and blurs the differences between a candid product review and a paid pop-up ad. These kinds of ambiguous messages make ads appear like editorials and are common in other forms of media as well, but computer technology streamlines and accelerates them, blurring the message further.

ONLINE POLITICS

In the run-up to the 2008 presidential election in the United States, most candidates set up personal MySpace pages, wherein the young and hard-to-reach users of this large social networking site could read the candidates' personal blogs, see their photos and videos, link to their sites, and learn more about their views on various issues—even donate funds to their campaigns. In the spring of 2007, MySpace introduced a special section called the Impact Channel, specifically dedicated to politics and the presidential election. It was intended to be an online version of a town square—a collection of political links to help the more than fifty million monthly MySpace users learn about the candidates by using the same approach they use to learn about their MySpace "friends." The video downloading site YouTube also became an important campaigning force in the 2008 election: presidential candidates debated on CNN in response to questions from YouTube users. With YouTube and similar websites, anyone can draw attention to anything they wish, whether it's a political misstep caught on a cell phone camera or a politician's damaging comment taken out of context.

A Pew Internet & American Life survey found that nearly a quarter of Americans reported that they regularly learned about the 2008 presidential campaign from the Internet, which is nearly twice the percentage at a comparable point in the 2004 campaign. Thanks to a greater number of cable stations, television viewers have more options, so TV campaign ads have a lower impact than in previous years, and Internet strategies are now major components of any political campaign. Emails, text messages, blogs, and interactive websites all help the political parties organize their campaigns, raise funds, and get out the vote more effectively than with traditional methods. The Connecticut research firm PQ Media estimated that $80 million would be spent on Internet ads for the 2008 presidential election.

The Web has proved to be particularly effective in raising political funds from young users. In his 2004 campaign, Senator John Kerry raised 80 percent of his campaign funds from eighteen- to thirty-four-year-olds through the Internet. The Web has also empowered individual citizens to get involved in the political arena—nearly anyone who wishes to have a voice can create a blog and reach out to other voters.

The Internet also safeguards the political process by adding greater transparency and accountability. Bloggers uncovered the forged documents relating to President Bush's military career that led Dan Rather to publicly apologize for his false CBS News story. Disinformation from anonymous Web sources can do considerable political damage.

When we think about or discuss political issues, a variety of brain regions become activated—from the emotional centers in the amygdala, where we consider the consequences of election outcomes, to the frontal areas, which organize our arguments. Political matters can stir up highly stimulating debates, which can trigger multiple neural circuits, whether you're liberal or conservative, a Democrat or a Republican.

Previous research on cognitive styles has supported the personality stereotypes of conservatives, who crave structure and order, and liberals, who better tolerate change and uncertainty. More recently, neuroscientists have explored actual brain function responses according to political affiliation. Professor David Amodio and associates at New York University used computer tests to assess how people naturally respond when confronted with tough choices, and the results suggest that political attitudes may be hard-wired in the brain. The investigators found that compared with conservatives, liberals demonstrated significantly greater activity in the anterior cingulate cortex, which monitors self-regulation during conflict. These results may reflect the mental flexibility of liberals and mental rigidity of conservatives. By contrast, a conservative might argue that these results should be expected from the illogical, soft, liberal brain compared with the logical and steadfast conservative one.

As the Internet and television bombard our brains with political messages, neuroscientists have identified the brain regions that control and monitor our responses. Even if we think we're not paying attention to the images or campaign messages, merely viewing the face of a candidate triggers a hard-wired response.

At UCLA, functional MRI scans demonstrated highly significant brain activation when research volunteers viewed the faces of presidential candidates. The pattern of their responses was driven by their political affiliations—their party's candidate triggered positive reactions, while the opposing candidate triggered negative ones. Viewing the faces of these political figures activated the neural circuitry that controls cognition in the dorsolateral (upper- and outer-facing) prefrontal cortex, as well as emotion in the insula and anterior temporal lobes. Thanks to the digital age, these neural circuits are now triggered by online politics in the same way they used to be stimulated by watching a political debate.

UPLOADING YOUR IBRAIN

Well beyond politics, shopping, and information gathering, the seemingly infinite number of activities on the Internet can be both empowering and anxiety-provoking. Many people experience a sense of freedom from knowing they can access whatever their heart desires, but others feel challenged by having too much to choose from. They may feel paralyzed and unable to make a decision.

There is no doubt that the digital age is growing more complex every day, and our brains, especially the younger ones born into a technology-saturated world, are continuing to evolve. To stay on top of our game and remain informed and involved in social, political, and economic trends, we all need to speak the same language—online and off—as well as have the skills to communicate face to face and use our human instincts to guide us.

Six

BRAIN EVOLUTION:
Where Do You Stand Now?

> *Man is still the most extraordinary*
> *computer of all.*
>
> John F. Kennedy

Reaching a balance wherein we use technology effectively while remaining connected to others in a personal way will help bridge the brain gap between Digital Immigrants and Digital Natives. However, we first need to determine our various strengths and weaknesses in the areas of our lives that we can control. Although many of us may have a general idea of where we may need help—text messaging, email etiquette, eye contact during conversations, or perhaps multitasking skills—the following assessments and the resulting scores will help you know where to begin training your brain. Not all baby boomers need to upgrade their digital skills; in fact, many could use a little help with their interpersonal relationships. Similarly, many young adults lag behind their tech-savvy elders in some areas of technology.

HUMAN CONTACT SKILLS

In our UCLA studies, we have used self-assessment questionnaires like those that follow, and we have found that research participants' self-rating scores correlate significantly with their brain function measures on actual PET and MRI scans. When answering the following questions, think of your work, school, and casual encounters as well as those involving close family members and friends. These questions will assess various nontechnological areas that have an impact on how well you function in society.

Nonverbal Communication Skills

Technology's lure can distract us and limit the time we spend in traditional social encounters. With too much tech involvement, our nonverbal communication skills, such as body language expression and interpretation, tend to suffer. These subtle signals actually say a great deal to others during a face-to-face conversation.

Answer the following questions by circling a number between 1 and 7 that best reflects how you judge your nonverbal communication skills. Afterward, tally your results to guide you on where you might wish to improve.

	Usually		Sometimes			Rarely	
When someone talks with you, do you find it difficult to maintain eye contact?	1	2	3	4	5	6	7
Do you have trouble interpreting the mood or meaning expressed by someone else's body language (tightly crossed arms, downward gaze, etc.)?	1	2	3	4	5	6	7
Do other people have difficulty interpreting your mood?	1	2	3	4	5	6	7
Do people comment that you seem distant, or often ask if something's wrong?	1	2	3	4	5	6	7
Do you feel uncomfortable when close friends or relatives hug or kiss you?	1	2	3	4	5	6	7
Do you feel uncomfortable meeting new people and shaking their hands?	1	2	3	4	5	6	7
Total score:_____							

If your total score was 36 or above, then you are probably quite comfortable expressing and interpreting nonverbal communication. If you scored 5 or below on any individual item, or if your total is between 18 and 35, the exercises and suggestions in the "Body Language" section in Chapter 7 may help you improve these skills. If you scored below 18, you should definitely practice the strategies in Chapter 7 and perhaps check Appendix 3 for other sources of assistance.

Self-Esteem

The ability to assert and express ourselves is an important social skill. People who feel confident are more comfortable speaking directly to others. Self-confidence also helps us to openly share our feelings and needs. Circle the number that best describes how you rate yourself in these areas.

	Usually		Sometimes			Rarely	
Do you have a hard time asking others for help or advice?	1	2	3	4	5	6	7
Do you have difficulty admitting when you make a mistake?	1	2	3	4	5	6	7
Is it hard for you to voice your opinion in a group, especially if you disagree with the consensus?	1	2	3	4	5	6	7
Do you ever agree to do things that you'd rather not do just because you don't want to disappoint people?	1	2	3	4	5	6	7
Do others complain that you are overly aggressive and critical?	1	2	3	4	5	6	7
Is it difficult for you to talk about your feelings honestly?	1	2	3	4	5	6	7

Total score:_____

If your tally was 36 or above, then you probably feel confident and are able to assert yourself with others. Scoring 5 or below on any one question, or scoring a total of between 18 and 35, suggests that the sections in Chapter 7 on "Effective Off-Line Communication" and "Building Self-Esteem" will be helpful to you. If you scored below 18, then in addition to practicing the self-confidence and assertiveness exercises in Chapter 7, see Appendix 3 for more resources providing help in this area.

Empathic Abilities and Listening Skills

Empathy—the ability to imaginatively see things from another person's perspective, understand the person's feelings, and convey that understanding back to the other person—serves as the social glue that keeps people connected. Fortunately, we can strengthen the neural circuitry that controls these abilities and improve our empathic skills at any age. Answer the following questions to get an idea of your current abilities.

	Usually		Sometimes			Rarely	
Do you ever lose interest when someone goes into detail about how they feel?	1	2	3	4	5	6	7
Is it hard to put someone else's needs and feelings ahead of your own?	1	2	3	4	5	6	7
Have you ever stopped being friends with someone rather than confront uncomfortable feelings they've instilled in you?	1	2	3	4	5	6	7
When a good friend or family member talks about their problems, do you feel disconnected or less close to them?	1	2	3	4	5	6	7
Do you feel uncomfortable talking about your true feelings to people you care about?	1	2	3	4	5	6	7

Total score:_____

Scoring 30 or above suggests that you are probably an attentive listener with good empathic skills. If you scored 5 or lower on any particular question, or if your total score was between 15 and 29, then the section on "Upgrading to Empathy 2.0" in Chapter 7 may help you in this area. If your total score was below 15, then be sure to try the exercises and strategies in Chapter 7 that apply to you. You may also be experiencing relationship problems, for which you might benefit from an individual or couples counselor or other professional advisor.

Multitasking and Attention

As we spend more time using technological devices we tend to increase our multitasking more and lessen our full attention on any one task. Answer the following questions to get an idea of your current abilities in these multitasking and attention areas.

	Usually		Sometimes			Rarely	
When interrupted from a task, do you have a hard time getting back to it?	1	2	3	4	5	6	7
When reading or listening to instructions, do you tend to miss important details?	1	2	3	4	5	6	7
If you get interrupted during a phone conversation and you tell the caller you'll call right back, do you ever forget?	1	2	3	4	5	6	7
Do you ever find yourself doing three or more tasks at once (e.g., checking email, talking on the phone, signing papers)?	1	2	3	4	5	6	7
Do you make mistakes, misplace things, or forget important information because of multitasking?	1	2	3	4	5	6	7

(continued)

	Usually		Sometimes		Rarely	
When trying to pay attention to something, are you easily distracted by other things going on around you?	1	2	3	4	5	6	7

Total score: _____

If your total score was 36 or greater, then multitasking and attention issues are probably not interfering with your daily life. If you scored 5 or below on any one question, or scored between 19 and 35 total, review the "Mastering Multitasking" and "Paying Attention" sections in Chapter 7. If you scored a total of 18 or below, review these sections but also see the Appendix for more resources that provide help in these areas.

Relaxation and Off-Line Creative Abilities

Spending hours online, answering emails on a hand-held device, or playing interactive video games can cause tension and anxiety, which affects our ability to relax. The hours we spend on the Internet clearly decrease the amount of time we have available to relax and be creative off line. Answer the following questions to get an idea of where you stand in these important areas.

	Rarely		Sometimes		Usually	
Do you find it easy to relax and unwind without using technology?	1	2	3	4	5	6	7
Do you take walks outside or make a point of spending time outdoors at parks, beaches, or other natural settings?	1	2	3	4	5	6	7
Do you enjoy painting, music, cooking, or other creative off-line pursuits?	1	2	3	4	5	6	7

(continued)

	Rarely		Sometimes		Usually	
Do you enjoy reading books for pleasure rather than online news, e-magazines, and blogs?	1	2	3 4	5	6	7
Do you set aside time to meditate, socialize, do yoga, exercise, read, or take part in other relaxing activities?	1	2	3 4	5	6	7
Total score:_____						

If your total score was 30 or greater, then it's likely you are able to relax and be creative off line. If you scored below 4 on any one item, or if your total score was between 15 and 29, you may wish to review the sections in Chapter 7 entitled "Turn Off the Gadgets and Turn to Yourself" and "Balancing the Creative Mind." I'd strongly encourage anyone scoring less than 15 to review these sections and try several of the strategies to help unwind and relax without technology.

Hi-Tech Addiction

Although you may not be suffering from Internet addiction or other technology addictions, there may still be devices or programs that pull you in that direction. Answer the following questions to gauge your tendency to technology addiction.

	Usually		Sometimes		Rarely	
Do you snap at others if they interrupt your online or other technology sessions?	1	2	3 4	5	6	7
Do you use technology to escape uncomfortable feelings or situations?	1	2	3 4	5	6	7

(continued)

	Usually		Sometimes		Rarely	
Does the time you spend at the computer, video gaming, or related activities interfere with your work or social life?	1	2	3	4	5	6	7
Are you ever defensive or secretive about your technology activities?	1	2	3	4	5	6	7
Do others complain about the time you spend on the Internet or using other technology?	1	2	3	4	5	6	7

Total score:_____

If your total score was 30 or above, then technology addiction is apparently not a problem for you. A total score between 15 and 25 may mean you have some addictive tendencies and might benefit from focusing on the "Technology Addiction" section in Chapter 7. A total score below 15 is consistent with an Internet addiction or related condition. In addition to Chapter 7, consult the Appendix for more helpful resources.

TECHNOLOGY SKILLS

Digital Immigrants or anyone who feels technologically challenged can catch up with their often younger colleagues and jump-start their technology skills. Chapter 8 provides a basic toolkit for using a variety of technological devices and programs. The following assessments will help guide you on where to get started.

Electronic Mail Expertise

Email has become a standard method of communication both at home and at the office. Most Digital Natives and Digital Immigrants use email as well as other electronic forms of communication. Some aspects of email, however, may still be unfamiliar to you. The following questions will gauge your level of email know-how.

	Usually		Sometimes			Rarely	
Do you avoid using email because it's challenging to master and you prefer other methods of communication?	1	2	3	4	5	6	7
Do you have difficulty organizing your email?	1	2	3	4	5	6	7
Do you think that responding to your work email cuts into your leisure time?	1	2	3	4	5	6	7
Are you uncomfortable using email for social communication?	1	2	3	4	5	6	7

Total score:_____

If your total sum was 24 or above, then you are reasonably experienced with email, but if you scored less than 6 on any individual item, consider reviewing the section on "Electronic Mail" in Chapter 8. A score between 12 and 23 means that you most likely could improve your email skills by using some of the strategies in Chapter 8. If your total score was 11 or below, you may be avoiding email altogether and will definitely benefit from the "Electronic Mail" section in Chapter 8.

Other Internet Communication Skills

Basic information on a variety of technologies, and strategies for dealing with them, are provided throughout Chapter 8, ranging from how to get started with instant messaging to tips on choosing the smart phone that's right for you. The following questionnaire will help guide you on where to begin.

If your total score is 60 or above, then you are at the skill level of many technology-proficient Digital Natives. A score between 30 and 60 rates you as moderately skilled, and if you scored below 30, then you definitely have a few things to learn about new technology.

Do you feel that you don't have a full grasp on how to use any of the following technologies and, as a result, find yourself avoiding them?

	Usually		Sometimes			Rarely	
	1	2	3	4	5	6	7
Text messaging	1	2	3	4	5	6	7
Instant messaging	1	2	3	4	5	6	7
Social networking	1	2	3	4	5	6	7
Blogging	1	2	3	4	5	6	7
Video conferencing	1	2	3	4	5	6	7
Internet phoning	1	2	3	4	5	6	7
Mobile phones	1	2	3	4	5	6	7
Hand-held devices	1	2	3	4	5	6	7
Internet search engines	1	2	3	4	5	6	7
Digital video recorders	1	2	3	4	5	6	7

Total score:_____

Although you may not need to perfect every new software program and hardware device, the technology toolkit in Chapter 8 provides a starting point that addresses these areas. I suggest beginning with those technologies where your assessment scores are low, as well as with any program or device that has piqued your interest.

Seven

RECONNECTING FACE TO FACE

Technology . . . The knack of so arranging the world that we don't have to experience it.

Max Frisch, architect and author

Sam, the owner of a classic jazz record label, was out playing golf with his buddy Phil. As they walked toward the sixth hole, Sam complained that his sales were slipping and he needed to find a way to put some pizzazz back into the business. Phil said his son, Jimmy, was an account exec at a hip young marketing firm that was supposed to be one of the best. Maybe they could help revive Sam's classic label—get the word out on the Internet and bring in a whole new generation of listeners. Sam called Jimmy, the twenty-four-year-old marketing guru, and the next day Jimmy emailed Sam some great ideas to promote his records. Sam excitedly set up a meeting with Jimmy to come in and pitch him.

Jimmy showed up with two equally young colleagues carrying laptops, an LCD projector, and a portable screen. Seeing the entourage and the gear, Sam suggested they move to the conference room, which was lined with gold records by acts these kids had never heard of. Sam offered them coffee, wanting to have a little chat, but the team demurred, wanting to get right down to the business of setting up their PowerPoint presentation. The lights went down, and the pitch was great.

Afterward, Sam was excited and had lots of questions. "Could they start immediately?" Jimmy said, "Yes." Sam asked, "Would you be in charge of the campaign?" Jimmy, packing up his computer, answered, "Yes." Sam said, "Will you want to use mainly radio spots?" Jimmy put on his jacket and said, "No. I'll email you a proposal." Sam got almost no eye contact from Jimmy or his team as they quickly picked up their gear and left.

(continued)

(continued)

A few days later Sam and Phil were on the golf course again. Phil asked, "How'd it go? Aren't those kids sharp?" Sam replied, "Yeah, they're very smart and the pitch was good." Phil asked, "So, when are they starting?" Sam said, "They're not. Look, Phil, Jimmy seems like a nice kid and all, but I need to be able to relate to the people I work with. Those kids could hardly look me in the eye. I'm telling you, Phil, they seemed like they were from another planet."

Twenty-somethings entering the workforce may know the latest shorthand for instant messaging. They may also have perfected their skills for gathering and manipulating vast amounts of information and images on the Internet. But all that solitary computer time leaves young brains less exposed to the vital stimulation of face-to-face social interaction. These young, tech-savvy Digital Natives often need to fine-tune their people skills. In fact, many of them could use a crash course in direct communication, including basic lessons in eye contact, empathic listening, and interpreting and responding to nonverbal cues during conversation.

Some young people have become challenged beyond fundamental social skills. They have gotten so isolated in their digital cocoons that they fall short in their essential knowledge of the practical world. In response to this educational need, many colleges have introduced courses on paying taxes, doing laundry, preparing meals, balancing a checkbook, and even dining out and using proper manners. The high-tech revolution has disrupted much of the basic life-skills learning that in prior generations would have taken place in almost any tight-knit family. Today, nuclear family members may still live under one roof, but they often substitute cyber interactions for traditional social exchanges with relatives and friends.

Research published in the February 2008 issue of the *Personality and Social Psychology Bulletin* indicates that everyday social contacts may boost brain power and cognitive abilities. In a study of more than thirty-five hundred people, University of Michigan psychologist Oscar Ybarra found that more time spent chatting with friends was associated with higher scores on memory tests. In a smaller sample of seventy-six college students, he found that volunteers who spent just ten minutes chatting with friends had better memory scores than those who spent the ten minutes reading or watching an episode of *Seinfeld*. The interactive, give-and-take qualities of everyday conversation appear to provide greater stimulation for our neural circuitry than mentally stimulating yet more passive activities like reading or watching a sitcom rerun.

Chronic Internet users risk other adverse psychological consequences. Symptoms of loneliness, confusion, anxiety, depression, fatigue, and addiction can emerge and further erode their social skills. The anonymous and isolated nature of online communication does not provide the feedback that reinforces direct human interaction. For example, an email message has a built-in delay before the response arrives, allowing the responder time to think about how to phrase the response and what style to convey it in. This delay can reinforce social inhibition.

By contrast, spontaneous face-to-face reactions from others help shape our own intuitive responses. Over time, these interactions create an accepted array of behavioral social norms, such as how to greet a stranger or co-worker or how to dine at an elegant dinner party. Corresponding brain neural circuitry controls each of these complex behaviors and social interactions.

Recent neuroscience points to pathways in the brain that are necessary to hone interpersonal skills, empathic abilities, and effective personal instincts. In Digital Natives who have been raised on technology, these interpersonal neural pathways are often left unstimulated and underdeveloped. However, electronic overexposure leading to altered neural pathways and waning social skills can happen at any age. Baby boomers and other Digital Immigrants also run the risk of becoming so immersed in the Internet and other new technologies that they experience a social and emotional distancing between themselves and their families and spouses.

THAT HUMAN FEELING

Just as the brain controls our ability to search online or answer email, it defines our humanity—our self-awareness, creativity, social intuition, and ability to experience empathy, trust, guilt, love, sorrow, and a range of complex emotions. Neuroscientists are discovering the underlying neural circuits that define these mental states and make us feel human. One key brain region, the insula, monitors the physiological state of the body and translates that state into the subjective experiences that drive our various actions, ranging from talking to eating to washing the car. The front part of the insula recasts a bodily sensation into a human emotional experience. For example, a caress becomes love, or an aroma transforms into lust.

Because the insula modulates basic functions like sex and eating, it was originally considered part of the primitive brain. However, neuroscientists Antonia Damasio, University of Southern California; John Allman, California Institute of Technology; and Arthur Craig, Barrow Neurological Institute in Arizona recently elucidated the insula's more complex role in the human experience.

Functional brain imaging studies show that sensations like smell, taste, touch, pain, and fatigue can activate the insula, which transforms feelings into more complex human experiences. Because it modulates cravings, the insula also triggers behaviors that can lead to addiction to drugs, alcohol, tobacco, sex, or the Internet. Smokers with insular damage may be able to give up cigarettes, but they will likely experience apathy, loss of libido, or other symptoms.

The insula lies deep to the brain's lateral surface and helps the brain sort out what is going on internally versus externally, which allows us to experience self-awareness and engage in social interactions. The insula is so finely tuned that it even anticipates experiences before they occur. When you walk outside on a cold winter day, your body braces for the experience in advance. The brain's insula also helps people to figure out if someone is telling the truth or lying. Memories of sad or excited feelings, the rejection you may feel when a friend doesn't return your phone messages, or the joy of listening to your favorite music—all these are insular responses.

The insula partners with other brain centers to contribute to the human experience. In the frontal lobe, the ventromedial prefrontal cortex controls moral decision making, while the orbitofrontal cortex helps us to make decisions about our future behavior. The anterior cingulate controls our ability to recognize facial expressions and intense emotions like rage and love. When we make mistakes, the anterior cingulate also fires, indicating its role in our experience of guilt and remorse.

A critical aspect of what defines human behavior is the ability to act appropriately in social situations and experience empathy. People who lack empathy to the extreme are sometimes called sociopaths. They tend to feel no guilt and to have no capacity for love. Sociopaths often break the law with no consideration of repercussions or consequences. They can be chronic liars, ostracized by society, and are frequently imprisoned.

Using functional MRI scans, neuroscientists have pinpointed a neural network that controls the ability to act correctly in a given situation. Dr. John King and his associates at University College in London scanned healthy volunteers while they played a video game designed to contrast aggressive to compassionate play. Whether the volunteers shot an aggressive humanoid or healed a wounded character, brain scans showed activation of two specific areas: the prefrontal cortex, often involved in complex reasoning, and the amygdala, a region that controls emotion. Both the aggressive and the compassionate actions were appropriate to the social context of the game, but the level of brain activation was clearly greater for the compassionate acts. Other research has found that injury to these brain regions results in socially inappropriate behavior.

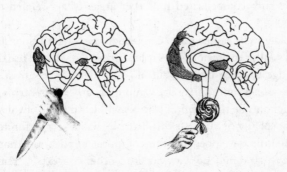

TECH-FREE TRAINING OF THE BRAIN

Brain scanning research has not only isolated the neural networks that define our humanity but also shown that we can take control and train our brains to refine our human behavior and social skills. Not only can our face-to-face communication talents improve with off-line training, but other complex mental abilities may sharpen as well. This type of training can include playing chess, learning a new language, taking up painting, or any number of methods of flexing the brain muscle in new and nontechnological ways.

Brain function generally does decline with age. It may take older people longer to learn new information and recall it later; however, some abilities improve with age: vocabulary, language skills, expert know-how, and emotional stability are among them. Thanks to life-long brain training as we age, our experiences are stored into neural circuits or mental templates that help us solve complex problems quickly and with little mental effort.

Dr. Arthur Kramer of University of Illinois and colleagues at the Massachusetts Institute of Technology studied older air traffic controllers and found that their reaction speeds, as well as their memory and attention abilities, were worse than those of younger co-workers. But when these researchers tested the study volunteers on realistic, complex, and fast-paced tasks, the more experienced traffic controllers outperformed their younger associates. They had the mental muscle to juggle multiple bits of information at a rapid pace. Their years of experience compensated for other areas of age-related cognitive decline.

An older well-trained brain can recognize new situations and problems as similar to previous ones it has already solved and use that prior knowledge to work out the current quandary. By contrast, the untrained younger mind may use a more linear, step-by-step approach. One might argue that Digital Natives will learn some of this complex problem solving online, but this learning will likely be limited to the mental skills developed for the computer application that is used repeatedly, and may not carry over to other contexts or real-life situations.

Driving is another example of how years of experience can improve judgment and save lives. According to the American Academy of Pediatrics, a sixteen-year-old driver is twenty times more likely to have a motor vehicle accident than any other more experienced driver, and two-thirds of the teenagers who die in car accidents are male. In fact, auto accidents are the leading cause of death for teenagers in America.

Years of face-to-face social interactions train mature people how to control their emotions, particularly feelings like impatience and anger that can lead to interpersonal conflicts. Professor Leanne Williams of the University of Sydney used functional MRI scanning to reveal the strengthened neural circuitry that older brains develop. Her team demonstrated that the medial frontal area of the brain—just behind the forehead—was more active in older volunteers than in younger ones when they experienced negative emotions.

Other research by Dr. Thomas Hess of North Carolina State University has demonstrated the so-called emotional intelligence of the socially experienced, mature brain. His group found that older research subjects were better able to judge character traits such as honesty, kindness, intelligence, or deception and to ignore irrelevant details about another person, in comparison with younger volunteers. Additional studies support the idea that the mature brain is more resilient than the younger brain and less prone to sadness and depression. Government scientists have found that older adults in their sixties and seventies report fewer sad days per month compared with people in their twenties.

Intervention studies using PET scans show that various forms of talk therapies can influence brain activation patterns. In depressed patients, psychotherapy stimulates certain brain regions known to control mood deep within the brain. Obsessive-compulsive patients who respond to therapy show decreased activity in the caudate nucleus and other deep brain areas. The psychological insights gained from discussing personal thoughts, feelings, and problems with a trained therapist can activate additional brain regions that control thinking and problem solving (frontal lobe) as well as memory and emotions (temporal lobe). All these psychotherapeutic interventions

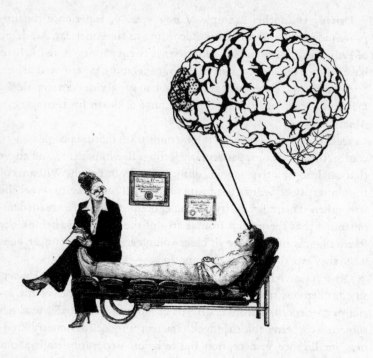

involve language and face-to-face contact, which contrasts greatly to the brain stimulation that comes from exposure to a computer or video screen.

Off-line brain training may counteract many negative consequences of extensive time online, particularly the neglect of a healthy lifestyle. Chronic Internet and technology users generally exercise less, gain weight, and experience more stress related to multitasking compared with people who rarely use technology. Our UCLA group studied what happens when research subjects, instead of constantly manipulating hand-held devices and computers, adopt a healthier lifestyle. We recruited primarily middle-aged volunteers to follow a healthy lifestyle program consisting of cardiovascular conditioning, memory exercises, relaxation techniques, and a healthy brain diet. After just two weeks, we found significant improvements in memory scores, as well as dramatic changes on their PET scans, demonstrating increased mental efficiency in the front part of the brain, which controls short-term memory and complex reasoning.

SOCIAL SKILLS 101

As the lure of technology distracts people of all ages from their usual personal interactions, their neural circuitry changes and everyday social skills begin to decline. The extent of these adaptations varies, depending on an individual's previous experience, amount of time online, and other influences. New technology has brought us remarkable advances, and the challenge is to take advantage of the technology without letting it take over other important aspects of our personalities. By identifying areas in our lives where off-line brain training can counteract the impact of digital stimulation on our mind's neural pathways, we can take control of how our brains adapt to new technology.

On the basis of your results from the assessments in Chapter 6, you will have a sense of what lifestyle areas might need fine tuning in order to rewire your brain accordingly. To get started, consider the following general tips for enhancing relationship skills and keeping technology overload at bay:

- Cut back on the amount of time you spend using all types of technology. Keep track of how much leisure time you spend answering email, talking on your cell phone, text messaging, watching television, or anything that does not involve face-to-face interactive contact with others each day. Add up the total time, and begin decreasing that amount by 10 to 20 percent at intervals that feel comfortable for you.

- As you begin reducing your time on one technological device, take care not to substitute it with another.

- Make a conscious effort to spend more time with people you care about. A fun, relaxing, fulfilling social life will reinforce your tendency to stay connected off line.

- Try to schedule regular family dinners. Recent research (Chapter 5) has found that teenagers who have a chance to reflect on their day with their families are less likely to abuse drugs, become violent, or engage in other high-risk behaviors.

- In addition to working on people skills, adopt other healthy lifestyle habits. Keep your brain fit with off-line mental aerobics, eat a healthy diet, get regular cardiovascular conditioning, and practice stress reduction techniques.

Body Language

Spending hours in front of the computer can atrophy the brain's neural circuitry that controls recognition and interpretation of nonverbal communication—skills that are essential for both personal and professional success. Some studies suggest that these nonverbal signals constitute a higher proportion of what we communicate to other people than the actual words we speak.

Functional MRI studies at UCLA have found that when volunteers focus their attention on another person's facial expression and tone of voice, brain activity in the medial prefrontal cortex increases. This brain region is critical for understanding other people's intentions.

What we say through a grimace, smile, tone of voice, casual touch, or rigid posture conveys a considerable amount of information to others. Many people are skilled in recognizing the more subtle gestures that others are telegraphing to them; however, even they can improve their ability to pick up on nonverbal cues.

To increase awareness of these important communication skills, first consider the many different ways we converse nonverbally:

- *General body language.* Styles of waving, nodding, pointing, and other gestures vary among cultures. How we stand, walk, cross our legs, or get up from a chair can let others know our mood, attitude, and energy level.

- *Facial expressions.* A grin, a frown, or even a perplexed look tells a lot about what an individual is feeling or attempting to convey. These expressions tend to be universal, regardless of culture or geography.

- *Eye contact.* Eye gaze can communicate a range of emotions, such as anger, passion, or sadness. Internet searches and email communications often make it harder to maintain direct eye contact when conversing with others because our eyes get used to flitting across a computer monitor while scanning for relevant information instead of looking straight at someone. Looking into someone's eyes while the person is speaking usually tells that person you are interested in what he or she has to say. Looking around the room or over the other person's shoulder suggests the opposite. Too much or too intense eye contact, however, can sometimes feel intrusive or seductive.

- *Touch.* Just as eye contact conveys a message, so does physical contact. Placing a palm on someone's shoulder can be reassuring; using both hands for a handshake can convey warmth. Such gestures are often culture based—many Europeans kiss both cheeks when greeting each other.

- *Appearance.* People often express themselves by the way they dress and wear their hair. Someone who is disheveled or sloppy could be interpreted as depressed or confused. Giving some thought to what you wear and dressing appropriately for an occasion is an effective way to convey your personality.

The exercise in the sidebar can improve your nonverbal communication skills.

NONVERBAL COMMUNICATION EXERCISE

Part 1. Think of something emotionally important that you can share with your spouse, friend, or close relative. It may involve a recent event, uncomfortable situation, or anything that has meaning to you or to the other person. Practice delivering your speech alone while facing a mirror. At first, make an effort to use *no* body language—just stand or sit still in a chair with your face expressionless and hands at your side. Do this for a few minutes.

(continued)

Part 2. Redeliver the speech to the mirror, but this time use facial expressions and your body to emphasize your delivery. Pay attention to your vocal tone, and be sure to emphasize important points.

Part 3. After finishing Parts 1 and 2, ask yourself the following questions:

How did you feel emotionally during the two parts of the exercise?

How much, if at all, did the clarity and impact of your message vary when you used body language?

What aspects of nonverbal communication—facial expressions, hand gestures, etc.—might you wish to improve?

This exercise can be varied by doing it with another person. Rehearse the message with someone you know well, who can give you feedback. Also, try swapping roles and have your friend or mate try to fine-tune his or her nonverbal skills while you act as the mirror and give feedback. For example, I might tell my wife—with a flat facial expression and in a monotone—how much I appreciate her kindness toward my parents. When I express this appreciation with a sincere, loving tone and a gentle touch on the shoulder, she smiles back at me sweetly and gives me a hug. When it's her turn, she may tell me how angry she is about my forgetting to take out the trash barrels—and I may actually prefer the version without the body language.

In addition to increasing your awareness of nonverbal communications and practicing the above exercise, keep in mind the following strategies to fine-tune your knack for communicating through body language:

• *Translate nonverbal cues.* When noticing another person's nonverbal expressions, think about what that person appears to be saying with body language. If you are not sure, ask. For example, if a friend seems fidgety and agitated and isn't making eye contact while you talk about a recent event, you might say, "I'm sensing that you aren't interested in what I'm saying, or maybe you're upset with me about something." This can give your friend the opportunity to open up about an incident you might not know about, or perhaps tell you something else that is upsetting your friend.

- *Look for nonverbal/verbal disconnects.* When we are conflicted about situations in our lives but unaware of those conflicts, we tend to nonverbally convey what we are *really* feeling. A clue to these hidden agendas is when someone's body language does not fit what the person is actually saying. Giggling when delivering sad news, and frowning when talking about presumably happy events, are examples of a feeling or attitude that is being suggested indirectly, under the surface. Pointing out the incongruent messages in a sensitive way may help the other person recognize the conflict.

- *Take in the big picture.* Body language, like all forms of communication, has a context. An affectionate kiss on the cheek may be heartfelt and sincere but out of place at a business meeting. Make sure that your nonverbal signals appropriately fit the particular situation.

Effective Off-Line Communication

The feelings associated with a specific conversation often dictate the method we choose for communicating, and as technology advances, the number of communication choices increases. During the past century, young people tended to use the telephone most often while courting and dating. Now teenagers gossip and flirt mostly by text messaging, video chatting, and IM'ing.

If we anticipate an uncomfortable outcome to a conversation, we may tend to avoid face-to-face interactions by choosing the least direct approach. For example, breaking up with someone is less confrontational when done via email, although much less personal and perhaps more hurtful. Dr. Adam Joinson of the Institute of Educational Technology in the United Kingdom found that when volunteers engaged in a high-risk communication like asking someone for a date, or the boss for a pay raise, they preferred to use the Internet more than direct communication. Joinson noted that the Internet helps people feel a greater sense of control during difficult interactions.

Ironically, the anonymity of Web communication allows some people to communicate more personal and intimate information than they might in person. Shy people often feel protected by email, but as they continue to shield their feelings from direct social contact, their ability to assert themselves with others may diminish over time. Although extroverts generally use online communications to expand their social networks, introverts can become more isolated with it. They may be avoiding the stress of direct human contact; however, they can end up feeling even more isolated and lonely.

It's possible to learn effective ways to directly express feelings and assert one's needs while still respecting the rights and needs of others. Assertiveness is not the same as aggressiveness, in which one does not usually respect the other person's point of view. Aggressive individuals can be manipulative and pushy, and they often force other people to give in to what they want.

By contrast, passive individuals are frequently ineffective in communicating their needs to others, usually because they fear conflict. While watching other people get what they want, passive individuals often experience resentment, anxiety, and depression. Sometimes they become passive-aggressive and retaliate indirectly. Rather than expressing their negative feelings directly, they may show up late for a meeting or make indirect negative comments.

Learning to be assertive—a skill that is often underdeveloped in Digital Natives—reduces stress and enhances everyday life. Assertive people are more likely to get their needs met and to free themselves of anxiety and resentment. They are often recognized as leaders and are admired and respected by others. They tend to be effective communicators and have satisfying relationships.

If face-to-face communication is a challenge for you, try some of the following strategies:

- *Face your fears.* Assert yourself by talking directly about your feelings. Keep in mind that expressing feelings does not necessarily lead to conflict. Imagine some of the outcomes that you fear might happen if you were to talk directly about your feelings: perhaps anger, disapproval, retaliation, frustration, or guilt. Whatever those outcomes may be, in the long run they are probably less toxic to your mental state than keeping the feelings bundled up inside.

- *Stick to the facts.* When asserting your needs, avoid making judgments or negative statements. Focus on the specifics. Instead of saying, "You're an insensitive idiot—you always manage to insult me," you might try something like this: "I didn't appreciate it when you criticized my cooking in front of our guests—you could have been more sensitive." Rather than using personal insults, the second approach focuses on the actual occurrence that triggered the feelings. Attempt to make your point without overstating the facts. Rather than saying, "You're always late, and now you've ruined my entire morning," try this: "When you're late for coffee, we have less time to catch up, and it makes me late to my next meeting." Sticking to the facts makes it easier to pinpoint the actual cause of a conflict.

- *Talk about yourself.* When conflicts begin, our first instinct is to blame the other person and focus on his or her faults. This generally incites the other person to retaliate, and the argument escalates. One way to break this cycle is to acknowledge your own contribution to the problem, which may help the other person admit a role in the conflict as well. For example, instead of saying, "I hate it when you slam the door, you big jerk!" you might try this: "I'd appreciate it if you wouldn't slam the door. It startles me and makes me nervous." This approach makes your feelings clear without attacking the other person, who will be less inclined to get defensive and argue, and more likely to hear your concerns and respond appropriately.

- *Assert nonverbally.* Keep your body language in mind when asserting yourself. Use effective nonverbal cues—relaxed posture, steady eye contact—to enhance your verbal message.

- *Listen and respond.* When we assert our feelings, we usually have a specific message in mind. However, it is best to avoid repeating the message over and over. Listen to the other person's point of view, and respond honestly, instead of just pounding home your message. If you don't understand the other person's position, ask for clarification.

Sometimes it takes practice to differentiate between aggressive and assertive behaviors (see Sidebar). For example, an aggressive response when your teenager borrows your laptop computer without asking might be "You spoiled brat. You always take my things without asking and end up ruining them!" An assertive response would be "You know you are not supposed to use my computer without asking, and I may need to put a password on it to keep this from happening again." The passive response would be to let your teenager continue to use your computer, then go hide in your room and simmer with pent-up anger. This might eventually lead you to a passive-aggressive response, such as "accidentally" deleting your teenager's files on the computer later.

ASSERTIVENESS ROLE-PLAYING EXERCISE

Try this exercise with a friend or your spouse to get a better sense of the fine line between assertiveness and aggression.

1. For each of the following scenarios, jot down an aggressive response and an assertive one.

2. Ask a friend to role-play each scenario with you. Afterward, describe how each response made you feel. Ask your friend how he or she felt when you responded each way.

Scenario	Aggressive Response	Assertive Response
Your overly talkative neighbor chats you up at the mailbox, but you have things to do in your apartment.		
Your boss passes you over and gives a younger co-worker the promotion and raise you were expecting and deserve.		

Scenario	Aggressive Response	Assertive Response
You agree to a weekly tennis game, but your partner habitually cancels at the last minute.		

Building Self-Esteem

People often have difficulty being assertive when they have low self-esteem. Individuals with little confidence may feel that their needs are not worthy of another person's consideration, so they don't express them.

Dr. Adam Joinson and co-workers found that volunteers with low self-esteem were significantly more likely to choose email over face-to-face communication. Email helped these insecure people avoid the anxiety they experienced when forced to interact directly with others and read their nonverbal cues. Left unchecked, continued isolation and avoidance of personal interactions can leave the brain's social neural networks underdeveloped.

Low self-worth also can lead to the online disinhibition effect—people start sharing their very personal feelings, wishes, and thoughts. At other times, low self-worth manifests with rudeness and angry threats. Using email, blogs, and instant messaging, some people, particularly adolescents with low self-esteem, become cyber bullies, harassing their peers with humiliating comments and embarrassing photos. Brain imaging studies show that impulsively aggressive people demonstrate decreased activity in the anterior cingulate gyrus and the medial orbital frontal cortex—brain regions that normally inhibit such aggression. The Internet provides a physical distance between the bully and the victim, which can escalate the viciousness of the attacks. This is a particular problem in teens, who lack the impulse control that most adults have because of their fully developed frontal lobes. However, many of these bullies fear direct confrontation and back down when they meet their Internet victims face to face.

Chronic and isolated Internet users can build self-esteem by spending more time with others off line. Of course, poor self-esteem can stem from a variety of causes, including underlying personality disorders, depression, and genetic factors. Psychotherapy has helped many individuals build self-confidence. Setting reasonable goals and working to achieve them also can enhance feelings of self-worth. Also consider some of following strategies to help boost feelings of self-esteem:

- *Look for triggers.* Ask yourself whether your feelings of low self-worth are related to poor job or academic achievement, social awkwardness, or financial troubles. Pinpointing what makes you feel vulnerable will help you focus on building yourself up in those areas.

- *Reevaluate your goals.* You could be setting your expectations for achievement beyond your abilities (or anyone else's, for that matter). Reasonable goals are often achieved in small steps.

- *Make moral choices.* Many people are tempted throughout life to cut corners and overlook other people's needs or wishes. When we instead take actions based on our ethical beliefs, we feel better about ourselves; even if it means sacrificing some of our own wants and needs. Helping others and supporting causes we believe in usually makes us feel good and boosts our sense of self-worth.

- *Focus on your strengths.* Routinely remind yourself of your accomplishments and strengths, and avoid dwelling on any weaknesses you perceive you have.

Having positive expectations about the future often builds self-esteem. Neuroscientists at New York University recently pinpointed the brain regions that control optimism. Dr. Elizabeth Phelps and her associates performed functional MRI scans while volunteers thought about positive future events. The scientists found significant activation in an area behind the eyes, the anterior cingulate cortex, and the greater the degree of optimism, the more activity. Other studies have shown that depressed people, who typically see the cup half empty rather than half full, have particularly low activity in this region.

One of the most common reasons why people doubt their self-worth is that they are in a slump—they have experienced a series of setbacks that makes them quickly forget their long-term accomplishments. Try the following exercise to put those slumps in perspective.

SELF-ESTEEM EXERCISE

Think of a recent setback, mishap, or perceived failure that may be eroding your self-esteem, and describe it in the left-hand column below. To counter this negative experience, jot down three accomplishments that have made you feel good about yourself in the right column. The negative-esteem trigger may be something that you have done or something done by someone else to you. It could be losing an account at work, being disappointed by a friend, or even an argument with a loved one. For example, if you feel low self-esteem when out with friends because your ex just got remarried and you are still single, try reminding yourself of the great relationships you have had in the past and the comfort you get from the close friends in your life right now.

Setback	*Accomplishments*
_____	1. _____
_____	_____
_____	2. _____
_____	_____
_____	3. _____
_____	_____
_____	_____

Upgrading to Empathy 2.0

Spending hours playing video games or working at a computer does little to bolster our empathic skills. Neuroimaging studies have identified the specific brain circuitry that controls empathy. Although this circuitry varies according to a person's abilities, most of us can strengthen our empathic neural pathways and improve our skills through off-line training.

Dr. Laurie Carr and her associates at the UCLA Ahmanson-Lovelace Brain Mapping Center used functional MRI to observe brain activity while volunteers engaged in an empathic exercise of mirroring another person's facial expressions. The participants were shown pictures of six facial expressions—happiness, sadness, anger, surprise, disgust, and fear. When participants looked at these pictures, brain activation was observed in the insula, the oval-shaped brain region that translates our experiences into feelings. When the volunteers imitated the expressions, the brain stimulation was in the exact same area, but significantly greater.

In other studies, Dr. Tania Singer and colleagues at the Institute of Neurology at University College in London studied couples in love. In their experiments, one partner experienced a brief painful stimulus, such as a small electric shock, and then later observed the spouse *appear* to be experiencing the same brief pain. Whether or not the spouse actually got the electric shock, the original partner *believed* the spouse felt it, and it triggered the anterior cingulate and the insula—brain regions that scientists have pinpointed as defining our empathy and humanity.

Empathy characterizes our species and has given us a survival advantage. The empathy of pre-humans led them to band together, which helped them survive the adversities of their environment. They were more effective in fending off predators and nurturing their offspring as a group rather than on their own. The brain's empathy centers clearly provide an adaptive evolutionary advantage.

Having empathic role models and experiencing our own pain can help shape our understanding of other people's feelings. The high-tech revolution, however, often detracts from these abilities. Although the content of an email or a text message may contain empathic feelings, the quality of the message is dramatically different from when it is expressed in person.

Learning empathy involves mastering three essential skills:

1. *Recognizing feelings in others.* Both verbal and nonverbal expressions can convey what another person is feeling and experiencing. Unfortunately, we don't always recognize these expressions because we are distracted or self-absorbed. The section on reading body language earlier in this chapter can help you fine-tune your abil-

ity to recognize other people's feelings. Also, keep in mind that someone who is expressing intense emotions can often take longer to express the verbal message; so be patient.

2. *Learning to listen.* The best conversationalists are people who know how to listen well. This involves putting aside distractions—both external (email, text messaging) and internal (random thoughts, worries)—and truly focusing your attention. Think of the last time you tried to explain how great or terrible your day was while your friend glanced down at the phone and responded to a text message from someone else. Sometimes, if we are excited about a conversation, we may interrupt the speaker to toss in our own thoughts. Unfortunately, by doing this, we run the risk of distracting or frustrating the speaker, possibly causing the other person to stop expressing real feelings. Good listeners have self-control—not just over allowing their mind to wander, but also over interrupting.

ATTENTIVE LISTENING EXERCISE

In this exercise, one person talks about something important in his or her life while the other person listens for five minutes. The listener actively focuses on eye contact and avoids interrupting or coaxing. The listener also catches his or her wandering mind. After five minutes, the listener and speaker switch roles. When both have spoken and listened, they discuss what the other person talked about and what the experience felt like. Many people find that by simply listening attentively, they develop an almost immediate sense of empathy and understanding for the other person.

---

3. *Letting others know you understand.* It's one thing to grasp another person's point of view, but the true power of empathy comes from communicating that understanding back to the other person. Try restating what you perceive as the other person's perspective, using simple statements such as: "Let me make sure I understand you" or "Tell me if I have this right." Asking for additional detail will show that you're interested in knowing more about the other person's situation.

There are many levels of empathic expressions (see Sidebar). Becoming more aware of these different ways of responding will improve your empathic abilities and bring you closer to others. Strengthening your social relationships through improved empathic skills will counteract the isolation that is typical of the new high-tech, digital age.

LEVELS OF EMPATHIC RESPONSE

Review the following scenario and possible reactions. Your friend complains to you: "I can't stand my boyfriend—I have to leave him. I despise his poker buddies, who are always hanging around the house. I really do love him, but he seems way more into his friends than me. I'm just too scared to break it off."

There are many ways you can respond. Here are several possibilities:

Unsolicited advice. "You'd be crazy to leave him. I know he really cares about you. No relationship is perfect—we all have to put up with something." Here you're advising and lecturing your friend even though she didn't ask for it. She probably won't feel understood or supported by you. Those of us who feel a need to fix problems usually have trouble tolerating the anxiety of anybody's ongoing conflict, and we tend to advise rather than empathize.

Sharing your own experience. "I know exactly what you're going through. I had a similar problem with my ex-husband—all he ever talked about was his stupid job. I finally had to dump him." Although sharing your own experience with a friend can be helpful at times, in this situation the timing doesn't seem right, and the response ignores what was being said. It may have been more helpful to find out more about your friend's story and then *ask* if she would like to hear about a similar experience.

Mirroring. "I had no idea you felt that way. You really seem scared and upset about the situation. Tell me more about what's happening." With this more empathic response, you actually use your friend's word "scared" and mirror her emotions. You show her you are interested in how she feels by asking her to tell you more.

Mastering Multitasking

Even people who can assert themselves well and who have great empathic skills may be challenged by multitasking. With a multitude of gadgets to choose from, the technology revolution has created countless situations that tempt us to multitask. After all, we're perpetually drawn into using new devices, yet constantly distracted by them. You can download music to your iPod while taking part in a conference call, or chat on the cell phone with your spouse while you clear out your email inbox—just make sure your spouse doesn't hear the clicking of your keyboard, or you'll be busted for not paying full attention. Trust me; I know.

Some forms of multitasking can actually be helpful, such as the ability to talk on the phone and search online to answer a question relevant to your phone conversation. However, as you increase your technology skills you may have a harder time managing the urge to multitask. You might be more inclined to start text messaging or playing BrickBreaker during a boring meeting, which is not only rude and frowned upon but does little to develop the neural circuitry controlling your people skills.

Neuroscientists have identified some areas of the brain involved in multitasking and have shown that the brain can be less efficient when attention constantly shifts from one task to another (Chapter 4). Some people, particularly Digital Natives, have fine-tuned their multitasking skills to their advantage, but quite often the habit limits productivity and leads to stress, anxiety, and inefficiency. Multitasking can also impair memory ability, since it distracts us from paying full attention to what we need to learn and recall later. People with fine-tuned technology skills who tend to spend more time online can become increasingly distractible and even hyperactive, multitasking from moment to moment.

Sometimes low-tech solutions are the key. Technology writer Danny O'Brien polled his list of top "overprolific" achievers and found a common low-tech secret to their productivity. They all used some form of simple, often hand-written to-do list to keep track of their tasks, usually checking them off as they were completed. For some it was a plain pad of paper with a list of calls they had to make and other tasks. Some used a clipboard to itemize things they had to do. Still others let their email inbox serve that function, sometimes emailing themselves quick reminders.

The following strategies can help keep multitasking in check. The key is to manage new technology and control the power it wields, rather than let it control you.

- *List your priorities.* Jumping from gadget to gadget and task to task can lead to a sense of disorientation and anxiety. To help avoid this, list your tasks and give each task a priority level. Start by completing your highest priorities first, then work your way down the list.

- *Timing is everything.* It is easy to get caught up in a particular task only to find that you're way behind in others. To avoid this, allocate specific amounts of time for each thing you need to do. Review your daily schedule, and figure out when you might be able to fit in time-sensitive and high-priority tasks. If you know that there is a block of time in the afternoon for a high-priority task, you won't feel so much pressure to keep chipping away at it in the morning. By assigning reasonable time periods for tasks, you will avoid mental fatigue and be able to maintain more attention on each task.

- *Take power naps.* Harvard scientists have found that a thirty-minute nap helps us to refresh our multitasking neural pathways (Chapter 1). If you feel techno-brain burnout kicking in, even briefer naps can help reduce mental fatigue or digital fog.

- *Alternate your tasks.* Neuroscientists have shown that varying your chores throughout the day significantly reduces the mental stress associated with multitasking. If you have 120 emails in your inbox, don't try to answer all of them in one sitting. Knock off a portion of the correspondence, and then move on to word processing, returning phone calls, or other jobs before going back to email responses.

- *Pause before moving on.* If you need to interrupt a word processing session or other activity, be sure to take a moment to note where you stopped before moving on to the next task. This will save time and minimize anxiety when you return to the original task.

- *Reduce physical clutter.* Many people are unaware that a disorganized, overly cluttered work space or living space can cause tension and anxiety as well as heightened levels of the stress hormone cortisol. An efficient way to reduce the clutter around you is to schedule a regular time each day or week when you turn off your cell phone and other distractions and use this time to file or throw away unnecessary documents, magazines, and other items that might be cluttering your space. My wife spends a half hour each week decluttering a different room or closet in the house, and she says it makes her feel great.

- *Stick to a routine.* Thanks to 24/7 Internet availability, people immersed in new technology may find themselves working around the clock, never giving their brains time to rest. Create a daily routine that makes sense to you—balancing work *and* leisure—and stick to it.

- *Set limits.* When you find you can't avoid multitasking, try to set a time limit for it. For example, if you are in the midst of writing an important document and a colleague interrupts you, tell your co-worker that you have only ten minutes to deal with that issue, and politely excuse yourself after that. If your colleague requires more time, set another meeting at a time that works for both of you.

- *Slow down.* Completing task after task at breakneck speed may make you feel that you are getting more things done, but you actually run a greater risk of making errors and having to redo your work. In the end, it may take longer to get the jobs done. By slowing down your pace, you may increase your accuracy and efficiency.

- *Build multitasking capacity.* We cannot eliminate all multitasking, but we can strengthen the neural circuitry that allows us to multitask more efficiently. The most effective approach is to build your multitasking abilities gradually. Begin by doing just two tasks together; however, make sure to choose tasks you are good at. People can use new electronic brain games not only to strengthen multitasking skills but also to cross-train their brains—alternating from left to right brain tasks (see Chapter 8 and Appendix 3).

Paying Attention

When we minimize our multitasking, we generally improve our ability to pay attention. If you're engrossed in finding the best online price for a new SUV, you probably have no trouble focusing mental attention on that task. However, if you are also trying to answer incoming emails and instant messages, your brain's neural circuitry shifts into overdrive, and your attention abilities suffer. This type of attention splitting over time can lead to attention deficit disorders (see Chapter 4).

In addition to the strategies to help you minimize multitasking, consider the following tips to help focus your attention:

- *Mind your mental wandering.* If you are one of the many multitaskers who tend to lose focus on the job at hand, make a conscious effort to catch yourself when your mind drifts off. Immediately bring yourself back to the current task. This type of self-monitoring can be very helpful to keep you from daydreaming or letting your attention drift. Some of the exercises on relaxation and meditation in the following section can also strengthen this skill.

- *Consciously engage your mind.* One reason many of us forget people's names is that we fail to focus and pay attention when we first meet the person. Paying attention often requires effort. If someone is talking about a topic that might not normally interest you, make an effort to find something noteworthy about it, perhaps a detail that has personal meaning. If the speaker is discussing next year's federal budget cuts, you could give it personal meaning by considering whether or not that particular cut will affect your own pocketbook.

- *Choose interesting tasks whenever possible.* When you have the opportunity, opt for activities you like. Heightened interest enhances attention. If you are bored with a particular task, knowing that the next one will be more interesting may be enough to push you through the duller activity.

- *Minimize distractions.* You may be completely engrossed in reading a Charles Dickens novel, but if a pickup truck crashes into your living room, you'll likely need to put the book down. Of course, most distractions are less dramatic, but making an effort to minimize them helps you focus on what you're doing. Telephones, computers, and hand-held devices are best kept out of earshot. If you're word processing, closing your email will eliminate the distraction of incoming messages. A comfortable seat and thermostat setting will also help keep your attention focused.

- *Take frequent breaks.* The longer the stretch of time you dedicate to one particular task, the higher the likelihood that you will get distracted. Attending a three-hour lecture on recent theories in quantum mechanics might lead the average person to become distracted before the second hour. Knowing that a break is coming up often helps people continue paying attention. During those breaks, be sure to stand up, stretch, make a phone call, or do anything different from the previous activity to refresh your mind so you can return your focus to the prior task when your break ends.

- *Consider medical intervention.* If paying attention and staying on task is truly a challenge for you, review the diagnostic criteria for attention deficit disorder (Chapter 4) and consider whether you might wish to consult a physician about your symptoms.

Turning Off the Gadgets and Turning to Yourself

Many of the strategies that help us focus attention also help us reduce stress and enhance our ability to relax. Whether you're struggling on your laptop to finish a report during lunch or stressed out by having to keep up with massive amounts of emails when you get home from the office, it is likely that technology is distracting you from your awareness of what is going on in your offline, nonwork world. In order to effectively reconnect off line with others, we also need to get a better sense of ourselves—how we actually feel physically and emotionally—and that can be hard to do if we are perpetually hooked up to one of our many electronic gadgets.

Mindful awareness or the relaxation response can be defined as the ability to turn off all technology and become more attuned to our present surroundings, feelings, inner thoughts, and physical state. Developing this skill not only helps us reduce stress but also helps us to listen and communicate better, which often improves our relationships.

Meditation is a popular method for achieving the relaxation response. Systematic electroencephalographic brain studies performed during meditation have demonstrated significant brain wave changes during and after a meditation session. Dr. Richard Davidson's team at the University of Wisconsin in Madison consulted with the Dalai Lama to recruit Buddhist monks, many of whom had ten thousand or more hours of meditation under their belts, and compared them with a control group of students with minimal meditation experience. The neuroscientists used functional MRI scanning to study brain activation patterns in these volunteers while they meditated, and found that the monks showed significantly increased activation in the neural connections between the frontal lobe and the amygdala, brain networks that control empathy and feelings of maternal love. The more years of meditation experience a monk had, the stronger the activation patterns. It was as if the years of meditation had strengthened the brain connections between thinking (frontal lobe) and feeling (amygdala).

Meditation, yoga, self-hypnosis, and other methods of achieving the relaxation response not only create a feeling of calm but also alter our physiology—heart and breathing rates decrease, blood pressure lowers, muscles relax, and immune function improves. These techniques often focus on a mantra or a repeated sound that works like a hypnotic rhythm. Meditators also learn to observe and let go of habitual mental chatter and to avoid focusing on distracting body sensations.

When Digital Natives and Immigrants find that their online and other technology activities have distracted them from focusing their attention on their present physical and mental states, they can benefit from regular daily mindfulness exercises. You don't have to become a Tibetan monk to reap the benefits of these exercises. I suggest starting with five-minute sessions twice a day and increasing the duration of the exercises as needed.

In these exercises, the aim is to keep the mind focused on one thought or mantra from moment to moment, relax physically and mentally, and learn to let go of extraneous mental chatter, allowing other thoughts to simply flow into and out of your mind. The sidebar provides three examples of stress-release breaks you can take throughout the day, designed to bring about this relaxation response.

The objective is simply to train yourself to take a rest from your usual technology-driven mental activity. When practiced on a daily basis, these kinds of exercises will likely improve your state of mind, general health, and possibly even your life expectancy.

STRESS-RELEASE BREAKS

Meditation Sampler

First choose a mantra, which can be any sound, word, or phrase that comforts you (e.g., "love," "om," "peace"). Sit in a comfortable chair or cross-legged on the floor, and rest your hands on the upper part of your thighs, palms up. Close your eyes and breathe slowly and naturally, allowing the muscles throughout your body to relax. With each exhalation, repeat your mantra silently to yourself, and try to focus your attention on both your mantra and your breathing. If outside thoughts come into your mind, let them pass through and out, and return your attention to your breathing and your mantra. After about five minutes, open your eyes and rest quietly so you can ease yourself back into the day.

Muscle-Group Relaxation Exercise

By paying attention to how you feel when your muscles relax, you can spread a sensation of relaxation throughout your body. The process is calming both physically and mentally.

Lie down or sit in a comfortable position. Breathe through your nose slowly, regularly, and deeply. Close your eyes and focus your attention on your forehead. Imagine releasing all the muscle tension there. Notice the sensation of the muscles relaxing as you bring your attention to your facial muscles and release the tension there. Let that relaxed feeling extend through your cheeks and jaw. Slowly continue this process, moving your focus down your neck and shoulders, releasing any tension and continuing to move systematically down your body to your arms, hands, abdomen, back, hips, legs, and toes. Continue to breathe deeply and slowly while relaxing all the muscle groups throughout your body. Spend a few minutes breathing in this relaxed state.

Picture Yourself There

While sitting quietly, close your eyes and think of a calming and memorable place you have visited, perhaps a favorite vacation spot or anywhere that makes you feel relaxed—a meadow, beach, or desert setting. Take a moment to recall the details of the locale, and focus on those details as you try to reexperience the feelings you had when you were in that setting. While breathing deeply and slowly, imagine yourself there. Feel yourself relaxing, and notice the physical sensations—perhaps the wind blowing on your face, or your toes wiggling in the sand. Spend a few minutes in this calming mental place. Then go back to your regular day. Don't forget to open your eyes.

Balancing the Creative Mind with New Technology

The high-tech revolution has had an impact on another defining human trait—creativity: our ability to generate new ideas as well as effectively execute them. Washington University psychologist R. Keith Sawyer points out that creativity generally does not occur with an instantaneous flash of brilliance but tends to develop over time after a series of insights that build upon one another. It often involves hard work and collaboration with others. Many medical and scientific innovators agree that discoveries often stem from collaborations among several individuals with different backgrounds and specialties. One of the benefits of Internet technology is that it allows people to connect with like-minded people at any time, any place. Architects, musicians, visual artists, and writers who have worked with new digital technology have benefited from the innovative ways it provides them to pursue their creative abilities.

Some, however, believe that too much technological stimulation may hamper the imagination. By restricting your interactions only to individuals who share your point of view, you may limit your communication with others outside your field of expertise, which may hinder innovation.

It takes a diversity of experiences—not just staring at a computer screen—to spark imagination and help discover the random analogies that can trigger an original idea worth pursuing. In the digital age, face-to-face communication *and* technological communication are key to maintaining our creative instincts. To balance your creative mind with new technology, try some of the following strategies:

- *Pursue new interests.* Different brain regions manage various aspects of creativity. For example, the left hemisphere controls writing through Broca's area in the frontal lobe and regions in the parietal cortex that direct the details of movement needed to create letter forms. When we are more creative, as in painting or sculpting, the right hemisphere takes charge: the parietal and visual occipital regions help us perceive the necessary spatial patterns, and the frontal lobe integrates and orchestrates the action. To expand your mind and flex your brain, try out new creative areas, especially if they involve unfamiliar skills and recruit neural networks that you may have been neglecting.

- *Brainstorm.* Whether you're writing the great American novel or a poem for your parents' fiftieth wedding anniversary, reserve specific creative times when email and other technologies cannot distract you from brainstorming and bolstering your imagination. If you plan to work with others, set aside this time together.

- *Be patient.* You can't force inspiration. Innovative breakthroughs can come to you when you least expect them. If you intersperse the times you plan to sit down and be creative with other pursuits and recreation, you will allow your innovative neural networks to rest and recharge.

- *Carry an idea recorder.* Whether you prefer a note pad, tape recorder, or hand-held digital device, keep one handy for that brilliant idea that may come to mind out of the blue.

- *Alternate low-tech and high-tech strategies.* To maintain optimal balance in your creative life, try to incorporate a variety of approaches. For example, if you have experience in writing classical sonatas for the piano, try out your teenager's electric synthesizer and see if you become inspired to compose a techno-hit or perhaps just scare the neighbors.

HIGH-TECH ADDICTION

The lure of new, fun, and time-saving technology is hard for many to resist. This attraction becomes a challenge when the technology stops being a useful tool and instead becomes the center of your life. Any hi-tech device or program can become habit-forming, but Internet and electronic gaming addictions are already widespread problems. Since they are a relatively new phenomenon, most experts have little or no experience in how to effectively treat them.

Because computers are everywhere—at work, school, and home—complete abstinence is usually not an option. Alcoholics can stay out of bars, but if an Internet addict happens to work at a computer, falling off the wagon is just a click away. Successful interventions help addicts to use the Internet in moderation, similarly to the way people with eating disorders learn to eat normally and in moderation. Support groups, twelve-step programs, and various psychotherapeutic approaches that have worked for other addictions are proving to be effective in treating Internet and video gaming addictions as well. Treatment centers focusing on drug addiction and alcoholism are adapting their programs to help people who also suffer from technology addictions (Appendix 3).

Sometimes Web addicts have underlying depression or anxiety, and treatment with psychotherapy, medication, or both can help diminish the compulsive online activities. Some addicts also have obsessive-compulsive disorder, which may respond to behavior therapy or antidepressant medication. Studies have found that group treatments of Internet sex addicts are highly effective in improving quality of life and symptoms of depression.

Whether you're already a full-on addict, or you just think you are spending too much time on the computer and other gadgets and it might be turning into a habit, some of the following strategies may help you gain greater control over the psychological pull of new technology.

- To protect yourself from the lure of tempting websites, try using software programs that filter and monitor content. You can also keep your home and work computers in public areas to lower the temptation of spending hours alone online.

- Identify your personal triggers to obsessive behaviors, such as boredom, anxiety, loneliness or other feelings or situations. Knowing your triggers can help you anticipate and avoid them.

- Find alternative off-line activities. Pursue hobbies and sports, go to the movies, and try out new types of leisure activities.

- Don't go it alone—spend more time with people you care about. Most experts agree that kicking any unwanted habit or addiction is nearly impossible to accomplish without the help of friends and family.

- If you think the lure of technology is interfering with your work or social life, you may wish to consult a professional to determine whether you have any underlying psychological issues that might be driving you toward addiction. Anxiety, depression, and other psychiatric conditions can be effectively treated, and this often improves addictive symptoms.

- Look for local resources and support groups that focus on your addiction. A key intervention strategy is to get help from others who have experienced the same or similar addiction and have learned to control it.

MAINTAINING YOUR OFF-LINE CONNECTIONS

Socially challenged Digital Natives and tech-heavy Digital Immigrants can't expect to change overnight. With greater awareness of their human contact challenges, coupled with social skills practice, they may soon see improvement in themselves, and others may see it too.

Sometimes we are not really listening as others speak but instead are thinking about what we're going to say next while we wait for a break in the conversation. It's usually easier and less stressful to fully focus our attention on the person who is speaking and simply respond accordingly. Often, simply making an effort to be friendly in social situations will encourage others to approach us and will help us overcome any anxiety or fear we may harbor about interacting face to face.

The strategies mentioned throughout this chapter can definitely help us all connect better socially. However, over time, the constant pull of technology may draw some people back into their solitary technology habits. It may be a good idea to re-assess yourself from time to time, using the questionnaires in Chapter 6, and to revisit some of these exercises and strategies to keep improving your human contact skills.

Eight

THE TECHNOLOGY TOOLKIT

The illiterate of the twenty-first century will not be those who cannot read and write, but those who cannot learn, unlearn, and relearn.

Alvin Toffler, *author of* Future Shock

You remember your first day of law school as if it were yesterday, and now here you are—you've really arrived. You're an associate on the rise at the largest litigating law firm in New York. Not only did they give you the cool smart phone that even the partners get—it downloads email, syncs your calendar and contacts to your desktop, and even reminds you of important appointments—but now a company limo is driving you to represent the firm at a hugely important deposition of one of their clients. You feel like calling your mother to gloat, but the limo has already parked in front of an intimidating skyscraper with the other law firm's name etched above the giant doors.

You arrive at the conference room on the thirty-second floor, and although you are five minutes early, your client and the opposing lawyers, ten of them, are already waiting for you. You sit beside your client and introductions are made. You notice that all the other lawyers have their phones—some wafer-thin, silver, space-age version of a BlackBerry, sitting on the table besides their pads. You take out your phone, which suddenly seems huge and clumsy, and place it besides your papers. You actually hear snickers from a couple of lawyers across the table. You now *hate* your phone, which at that exact moment goes off loudly with Cher's "Do You Believe" ringtone. Damn! You forgot to shut off the ringer! You excuse the interruption and silence the call, which actually *was* your mother, calling to wish you luck.

For many of us, our hand-held and other technological devices become more than just tools to make our work more efficient. They actually become extensions of ourselves, evoking feelings of pride, embarrassment, and envy. Whether we're experts or novices in using modern technology, constant advances require us to continually update not just our devices but our abilities as well. The good news is that technology skills can be learned and new neural pathways can be created in our brains at any age.

As people over age fifty-five continue to stay in the workforce longer, they may find that upgrading their technology skills will help them mesh with a younger workforce and culture that's packing breakneck information-gathering, problem-solving, and multitasking abilities—talents that are hard-wired into their young brains from a very early age. Many people first become familiar with a particular computer application, such as email or word processing, at work but soon find it helpful in their personal lives as well.

The assessments in Chapter 6 may help guide you on where you wish to focus in this toolkit. In addition, the appendixes include a glossary of high-tech terminology (Appendix 1), text message shortcuts and emoticons (Appendix 2), and other resources (Appendix 3).

MAKING TECHNOLOGY CHOICES

When in the market for a new computer, we often make our choices in the same way we buy a sports car or decide on a vacation, based on feelings and not on thoughtful analysis. Although style and "bells and whistles" surely influence our computer-buying decisions, more important considerations might include how we plan to use the device, its processing speed, hard-drive capacity, amount of memory, type of operating system, and other practical options. To stay on top of the latest innovations, it's helpful to check updates from technology experts in trade magazines, newspaper columns, and blogs.

When we make decisions about which technology device to purchase, once again our brains look for familiar brands, like Apple or IBM, which fire up the insula and the anterior cingulate—neural circuits in the frontal lobe that control positive emotions. The insula also monitors unpleasant emotions, so overpriced devices will further tweak those brain regions.

Neuroscientists and economists have pinpointed other neural pathways that monitor our purchasing choices, such as a deep brain structure known as the dorsal striatum. Caltech investigator Ming Hsu and his colleagues used functional MRI scanning to study volunteers while they made decisions based on different levels of missing information or ambiguity. An example of an ambiguous choice might be what to wear to a gala event when you're missing important information, such as how formal it is, the weather forecast, the time of day, or whether it involves much walking.

When faced with ambiguous choices, the brain's amygdala and orbitofrontal cortex consistently fire up. The amygdala generally monitors situations requiring vigilance, and whether or not to trust another person. The orbitofrontal cortex integrates our emotions and logical thinking. The more information we have about a technology purchase, the easier it is on our brains to make a decision.

Deciding between a laptop or a desktop computer is becoming harder as laptops get more versatile and do pretty much everything that a desktop can do while remaining portable. However, for the same speed, memory, and options, desktops tend to be less expensive. Desktops are also ergonomically more comfortable to use, and generally have larger screens.

The speed of your computer will depend on the processor or CPU (central processing unit), which is essentially the brains of your computer. If most of your work involves basic programs, such as word processing or simple spreadsheets, you may not need the high-speed versions required for complicated programs or large data files. High-speed processors can get pricey, and often the best value lies in purchasing the computer that is one notch below the top of the line

Memory is the amount of information that your computer can store, and you can usually expand the amount of memory on your existing computer for a price. Random access memory (RAM) stores the programs that control the computer's operating system.

Your computer's hard drive stores your programs, files, photos, and other data. A hard drive with 100 GB may be all you need if you're not working with graphics, video, or audio, which require more memory—sometimes up to 500 GB. A newer technology called solid state drive (SDD), made of memory chips without moving parts, may eventually replace older and slower hard drives.

YOU'VE GOT EMAIL

Email allows us to quickly convey information, such as a meeting time and place, or a response to a request or question. The user is able to avoid playing phone tag and can instead get a message across efficiently. Email lets us send notes to multiple recipients at once, and makes it easy to store and retrieve the documents.

However, email does have its limitations. Because it's so easy to knock out a quick email, we often use it for exchanges that could be discussed in person, or at least on the phone. When people disagree on an issue, email is not always the most effective way to reach consensus. Without direct social contact, body language and subtle emotional expressions are lost. Also, the true meaning behind an email comment can be easily misinterpreted.

In a typical face-to-face conversation, we react from moment to moment to what the speaker is saying. The nonverbal cues of the listener often shape the speaker's delivery. If I notice the other person is reacting negatively to what I am saying, I might soften my tone in midstream. If the listener seems bored, I might pick up the pace or change my tack. Email does not allow this kind of real-time adjustment when we deliver a message. In this way, email tends to break down our natural tendency to monitor or adjust ourselves during conversations.

Regardless of your reason for sending the email, consider whether a phone call, handwritten note, or face-to-face meeting would make more sense. If you need to convey important and sensitive information—an announcement of an illness, a job promotion or demotion, or perhaps a proposal of marriage—email may not be the best approach.

Take Your Time on the Subject Line

Craft your subject line so it summarizes the message. Specificity and detail are essential. Rather than labeling your subject "Conference," try

"UCLA Research Conference 4/1/09." Be mindful that some recipients will view the message on a hand-held device, which limits the number of characters in a message. A subject line that reads "2009 Conference on New Research to Be Held on April Fool's Day at the University of California, Los Angeles Campus" might instead appear as "2009 Conference on New Rese," thus failing to convey the intended content. Sometimes it is enough to simply put your name in the subject line to let the receiver know that it's you writing and not to delete without reading.

Make It Clear and Get to the Point

Clarify your reason for writing up front. If possible, try to pack the main details and purpose of the email in the first sentence. Concise writing that gets to the point quickly will increase the likelihood that your email will be read in its entirety. For maximum impact, take the time to trim and focus your message.

Mind Your Form

The layout of your email is important, since reading text on a screen is generally more difficult than reading a hard copy. Keep your paragraphs short, and leave a space between them to increase visual clarity. Try using bullets to delineate points, and AVOID WRITING IN ALL CAPITAL LETTERS. Not only is it harder to read, but it makes it seem as if the writer is shouting. Here are a few other pointers to help make your form more effective:

- Don't go crazy with exclamation points!!!!! Save them for an occasional "Congratulations!"

- **𝔄𝔳𝔬𝔦𝔡 𝔣𝔞𝔫𝔠𝔶 𝔣𝔬𝔫𝔱𝔰. 𝔗𝔥𝔢𝔶 𝔞𝔯𝔢 𝔥𝔞𝔯𝔡 𝔱𝔬 𝔯𝔢𝔞𝔡 𝔞𝔫𝔡 𝔡𝔦𝔰𝔱𝔯𝔞𝔠𝔱𝔦𝔫𝔤.**

- Stick to black font color except for responses, when a blue tone may help differentiate individual correspondences in an email chain.

- Go easy on abbreviations and emoticons (Appendix 2), especially for business messages. "LOL" or "☺" doesn't usually play well when you are applying for a new management position.

Respond to All Items in a Message

Emails often include several action items and queries. To increase the efficiency of your electronic correspondence, make sure you address each of the items in your response. As Will Schwalbe and David Shipley wrote in their book *Send*, be sure to scroll down to read everything in the email before replying. If you respond to only two of several questions in the initial email, it requires the correspondent to come back and ask about the other items again.

Include the Message Thread

When you respond to an email, it's usually best to choose the "Reply" option. If you click "New Mail" instead, the thread or original mail will not be included in your response. Including the thread ensures that the recipient knows what your email is in response to and gives it a sense of context.

Send Only What's Fit to Print

Before clicking the Send button, remember that any email message can be used against you in a court of law. Some business consultants recommend a five-minute buffer period between composing an emotionally charged email message and sending it. That way you have a chance to get additional perspective and to consider whether you really want to send the message. Consider never sending emotionally charged messages that you cannot take back. An email is a written record that can be forwarded to others—and sometimes altered—to your discredit. Avoid emailing confidential material, sensitive information, and discriminatory remarks. Know that jokes, subtle messages, and intricate conversations work best in face-to-face or telephone chats.

Personal Notes Versus Mass Mailings

Although it can be efficient to send announcements or automatic replies to large groups, if you want recipients to respond, it's best to personalize your messages. Many people tend to quickly scan or simply delete "Dear Friends" messages, considering them another form of junk mail. The perception is that if it were really important, the sender would have personalized the message.

To CC or Not to CC

One annoying habit of some emailers is to include *all* the addresses of multiple recipients in the visible "cc" (carbon copy) list. Equally irritating is when someone responds to these mass mailers using the "Reply to All" option. Not all recipients on the "cc" list are interested in the details of your schedule, why you cannot attend a particular meeting, or why you can't bring the donuts to the Little League game.

Another drawback of listing multiple individuals in the "cc" box, especially if you are forwarding a message, is that it may publicize the email addresses of individuals who would rather remain private. If you don't want others to know who else is receiving a copy of the email, use "bcc" (blind carbon copy).

Attach with Care

With the attachment function, you can share manuscripts, images, videos, and other documents, but some attachments are too large to download and use up huge amounts of memory on the receiver's computer, as well as significantly slow it down. Compression programs can decrease the attachment size and make them more compact. If you must send large attachments, websites like Pando.com and YouSendIt.com offer free software that allows you to transmit larger attachments. If you can't avoid the snags of large attachments, try copying the files to a CD that can be hand delivered or sent by regular (snail) mail.

Beware of attachments from strangers—they may contain viruses that can disrupt your computer's operating system. *Never* open a suspicious attachment from an unknown sender with the message "The file you requested."

One way to avoid adding unnecessary attachments is to copy the relevant information from a word file and paste it directly into the email message. Appendix 1 (High-Tech Glossary) includes definitions and descriptions of some common types of attachments.

Keeping Up with Your Inbox

We usually use email because we want and expect quick responses. Most emailers generally anticipate a response within twenty-four hours. A delay of two days is excusable, but if you wait beyond that you're conveying disinterest to the sender. If you can't respond within a reasonable period, at least let the sender know that you received the message and will send a complete response later, perhaps even specifying when. Another approach is to use an automatic out-of-office reply that tells senders you are unavailable until a specific future date. Setting aside time each day to catch up on your email is a great way to keep your inbox from getting out of hand.

Hoarders Versus Deleters

Some people take pride in keeping their email inbox nearly empty, while others hold on to nearly every correspondence. It's a good idea to create file folders to store emails according to topic but to save only those emails you may need for future reference. The less complicated your filing system, the easier it will be to retrieve previous correspondence.

Managing Your Email Time

Many people have difficulty controlling the impulse to constantly catch up on email. This perpetual multitasking habit can at times become addictive. When we spend too much time in front of the computer or staring at our hand-helds to keep up with email, we often neglect other tasks and may become inefficient in our daily lives. If you think that you need to cut back on your habit of checking email and wish to have more control over the firing up of your dopamine and anterior cingulate circuits, which are the neural basis of any addiction, keep in mind the following strategies:

- Schedule daily email sessions, and try to limit them to an hour or less.

- If you can take care of an email response in a minute or two, deal with it and move on. Otherwise, save it for scheduled email sessions.

- Use an email filter to sift through and automatically isolate junk mail and spam.

Helping Others with Email Etiquette

Many friends and colleagues habitually send jokes, chain letters, long attachments, and other distractions during the workday, and sometimes it becomes necessary to let these people know, tactfully, that you would prefer not to receive such correspondence. A comedy-writer friend of mine was receiving up to twenty jokes a day from colleagues. He finally sent out a mass mailing to all his friends and colleagues saying that he was too busy to read their jokes and would they mind removing him from their email joke lists. It worked like a charm. The truth is, I've never emailed him anything ever again.

More Email No-No's

Chain letters are usually spam that add to email clutter and are better deleted than forwarded. Also, be cautious of responding to the "unsubscribing" option of an email newsletter. Your response confirms to the sender that you received the mail, which may generate additional unwanted mail or even viruses that can infect your computer's operating system. Either delete the note or use software that removes or quarantines unwanted mail, possible viruses, and other spam. Finally, if you send out an email in error, rather than sending a follow-up note with a "recall" message, you're probably better off sending a brief, more personal follow-up note clarifying the mistake. One way to minimize such errors is to enter the addressee in the "to" box *after* composing the note—that way you avoid accidentally and prematurely pressing the Send button.

INSTANT MESSAGING RIGHT NOW!

As the name implies, instant messaging, or IM'ing, is a program that allows users to send and receive messages instantly—much faster than regular email. Although originally designed for quick personal messages, this form of fast-lane messaging has made its way into the workplace. Many individuals in business settings prefer IM over telephone or email because it provides immediate information and feedback.

Many people let their guard down when they IM, believing that the text is wiped away automatically. They may criticize the boss, ridicule a co-worker, or tell an indiscreet joke. However, anyone in your IM discussion can save the conversation for future viewing or use, possibly against you. To protect your privacy, make sure that your IM screen name does not appear in any public areas, such as Internet directories or online community profiles.

It is also a good idea to avoid IM'ing any sensitive or personal information (e.g., credit card numbers, passwords). Avoid IM attachments, since viruses appear to penetrate IM firewalls more easily than email firewalls. If you have kids who use IM, you can use parental control programs to keep tabs on them and limit their IM activity.

If you use IM for both work and social communications, separate your business IM buddy list from your personal one. That will keep you organized and help you avoid bringing your social IM'ing to the office. If you do IM socially during work hours, keep it to a minimum and limit it to break times.

SEARCH ENGINES: BEYOND BASIC GOOGLE

With the growth of the Internet and the power of its search engines, people search online more than ever before. However, recent data indicate that many people are inefficient in the way they search on the Web.

One reason for our search inefficiencies is a tendency to multitask while online. Neuroscientists have found that if we are searching on Google while talking on the phone or distracted by some other activity, our switching of attention from one task to another fires up a brain region just behind the forehead in the anterior prefrontal cortex. This area lets us leave a task, even if it is incomplete, and get back to it later from where we left off. Interestingly, this is one of the last parts of the brain to develop in young children, as well as the first part to decline in older people, so it is no surprise that both young children and older adults find multitasking to be a challenge.

The following suggestions can help reduce multitasking challenges by fine-tuning your search engine skills.

- *Trim your keywords.* The AOL database showed that people often type in an entire Web address instead of just the necessary keywords. Save yourself a few keystrokes when you want to visit a site like www.eBay.com—drop the "www" and the ".com." Simply typing "eBay" will do just fine.

- *Home page search engine.* Make your favorite search engine the home page on your computer. That way you can go right to the search and avoid having to type "Yahoo," "Google," or your search engine of choice whenever you sign on.

- *Advanced searches.* These features allow you to fine-tune your searches. For example, if you wish to search *hit songs from the sixties,* you will get results for each word you entered—almost two million. By merely putting quotes around your search words or phrase, you will narrow your search significantly.

- *Online dictionary.* In the search box, just type the word "define" before a word, and you'll get a list of definitions for that word. Don't worry if you've never won any spelling bees—if needed, the program will suggest an alternative correct spelling for your word.

- *Calculator.* Type in a mathematical equation, and the engine will solve it. If you must quickly know the answer to the equation "$783 \div 22$," your result will appear in an instant: "35.5909090909."

- *Currency converter.* If you are flying to Paris for the weekend and are not sure whether you can afford a four-star hotel, type "650

Euros" into the search window to learn that you will need $950 per night. Bon voyage.

- *Phone book.* Forget about that heavy yellow phone book—simply type in the name of a business and its city or zip code to get the address and phone number. If you want to know the city for an area code, just type in the three numbers of the code. When you type in "212," not only do you learn that it is the area code for Manhattan, New York, but you get a link to a map of the area. If you want to learn the area code for Manhattan or any other city, type in "Manhattan" or the other city's name.

- *Almanac.* Whether you want to know the capital of Burkina Faso, West Africa (Ouagadougou) or the tallest building in Hong Kong (Two International Finance Center—1,362 feet), you can quickly find the information by typing the phrase into the search engine query box.

- *Get a library card.* These low-tech cards will provide you free access to the high-tech search engines at your local library. Once you get the card, you can log on from your home or office computer.

TEXT MESSAGING: SHORT AND SWEET

Text messaging is a way to send short written notes from one mobile phone or hand-held device to another. These messages are no longer than 160 characters and can be a convenient way to quickly transmit information without having to directly phone the other individual. Because these hand-held devices are often small and sometimes awkward to manipulate, keeping your text notes brief is key—rmbr 2 txt n shrthnd.

Some Digital Immigrants complain of joint stiffness and pain when they text on hand-held devices with tiny keypads. Stretching and flexing your hands and fingers before and after texting will reduce joint discomfort (see Sidebar). Newer touch-screen gadgets are more joint friendly. When using text shorthand, remember that one slip can change the meaning of a message. If your text reads: "I hv 2 go mob" (I have to go mobile), the meaning of your message is very different if you slip and add an extra letter, as in "I hv 2 go myob" (I have to go, mind your own business).

STRETCHES FOR HAND-HELD TENDINITIS

1. Make a fist with your right hand and hold it for five seconds, then gradually open your fist until your fingers are stretched out and splayed. Hold for five seconds. With your left hand, gently squeeze each finger of your right hand, pulsing slightly, to increase blood flow. Start with your thumb, and then move to your index finger and so forth. Switch hands when complete.

2. Place both hands together, fingertips touching. Extend and flex your fingers while maintaining fingertip contact. Keep your movements slow and deliberate and feel the stretch.

3. After completing both of the above stretches, gently shake your hands as if you are trying to air-dry them. This will further increase circulation and relax the joints and tendons.

Some people, especially young Digital Natives, have become habitual texters. They can't seem to stop sending and receiving text messages, whether at work, at school, or with friends. It's always best to avoid texting during meetings, meals, and other social interactions. Also, never drive and text at the same time. Washington, New York, and California are among the states that now slap hefty fines on anyone who gets a DWT (driving while texting), and many other states are enacting laws against using any hand-held devices, including cell phones, while behind the wheel.

The type of mobile unit you use will influence your effectiveness as a texter. Devices with larger keypads are easier to manipulate, as are those with a keyboard that includes a separate button for each letter of the alphabet versus a standard phone pad that shares letters on fewer keys. The latter requires repetitive pushes on buttons for many of the letters. To jump-start your texting agility, see Appendix 2 for a list of TM shortcuts. B4 2 long, u'l b txting 2.

MOBILE PHONES: SMALLER IS NOT ALWAYS BETTER

Many baby boomers remember when a high-tech phone meant push-buttons instead of a dial. Many also recall the first generation of mobile phones that weighed more than ten pounds and required a small suitcase to carry around. That has all changed, of course. Now, some phones are so small that they need to be kept away from young toddlers lest they swallow them whole.

Today, cell phones are everywhere—in executive boardrooms, elementary schools, even atop Mount Everest. A recent National Highway Traffic Safety Administration Study found that driving while holding a cell phone to your ear can increase your risk for an auto accident by at least 30 percent. Even speaking on a hands-free device divides your attention between your conversation and your driving. Recent studies have found that the attention impairments associated with using a cell phone while driving are comparable to those associated with driving while drunk.

Functional MRI studies have found that when we focus attention on a cell phone activity like making a call or text messaging, three brain regions—the left prefrontal cortex, the anterior cingulate, and the parietal lobe—work together to complete the task. But as we focus our attention on the phone call our brain often misses other important information coming in. Peripheral vision is particularly affected. While talking hands-free on the cell phone, I myself have passed by the usual freeway exit I take to get home. Now I keep my cell phone use to a minimum in the car.

Along with increased use of mobile phones, we are seeing more frequent breaches in cell phone etiquette. We're all routinely reminded to turn off the ring tones of our phones before movies or lectures begin. To ensure that you don't commit a cell phone faux pas, keep in mind the following:

- Many public places restrict cell phone use. As a rule, don't use your phone in hospitals, elevators, professional offices, libraries, museums, or restaurants.

- Making or accepting personal cell phone calls or text messages during a face-to-face business or social meeting is just plain rude. The rare exception is a medical or other emergency.

- While talking on your phone, make sure you are at least ten feet away from other people, and try to speak quietly so your conversation remains private and you do not disturb those nearby. Not everyone is interested in hearing the details of your business meeting or altercation with your mother.

- When beginning a cell phone conversation, say that you are on a mobile device and that you will get back to the other person if the cell signal cuts off. (Of course, this allows you the opportunity to just hang up if the conversation becomes painfully boring.)

A challenge for Digital Natives and Immigrants alike is deciding on which mobile device to buy next and what service provider to use. Before choosing a new device, consider whether your current provider (e.g., Sprint, AT&T) has a strong signal in your area. Some phone carriers now provide phone service that switches seamlessly between Internet Wi-Fi and regular cell phone networks. If you're a frequent traveler, international service may be an important option for you.

Internet and newspaper ads showing the latest hot phones may be appealing, but there is no substitute for picking up the actual phone at your local provider's store and testing it out yourself. Check out the digital display and make sure you can read it easily, even in dim light or outdoors in the sunlight.

Some Digital Immigrants prefer simple "retro" phones with only basic phone service and text messaging, but the down side is that you might end up having to purchase a second device to use as an organizer, and perhaps a third one as a pager. Soon you've run out of pocket or purse space to carry all these devices around. Many people today are opting for a combined all-in-one phone/organizer/hand-held Internet device, rather than separate units.

A MENU OF HAND-HELD DEVICES

When first introduced more than a decade ago, hand-held organizers, or PDAs (personal digital assistants), were relatively simple gadgets that included calendars, address books, and to-do lists. They essentially replaced the daily planner books that many of us kept in our pockets or purses. I liked my old daily planner—it was easy to use and handy, but it had drawbacks. If I lost it, all my critical information was gone. Also, I had to routinely recopy my addresses and phone numbers when I purchased my new planner for the coming year.

Electronic organizers solved many of those problems. Because the data could be downloaded to a desktop computer, I still had all my important information backed up in case the organizer was lost or stolen. Also, there was no need to recopy addresses and phone numbers to my new daily planner each year. If I purchased a newer PDA, I just uploaded the relevant information to the new device.

Today, some hand-helds have nearly all the functions of a desktop computer, entertainment center, and communications hub. Instead of being just address books and calendars, they are also phones, mini-computers, video cameras, portable email transmitters, and cameras. They can be used as voice recorders, walkie-talkies, GPS map systems, and MP3 players. And size does matter. As portable technology develops, these hand-held devices are becoming smaller, slimmer, and more efficient. If you're in the market for a new hand-held device, you'll want to keep in mind the following points:

- *Cost.* The more you spend, the more you usually get, but remember that the average life span of these devices is about a year or two, so it may not be wise to invest too much. If you use the device for work, you may be able to get it paid for by your employer or deduct part of the expense from your income taxes.

- *Size.* Larger devices tend to have more features, but hand-helds that are too large can be awkward to carry around. My first test is whether I can slip the device into the inside pocket of my suit coat. For women, it may need to fit into a small purse.

- *Power.* Gadgets with many features may use up a lot of battery life. Make sure that the one you choose has adequate battery life to get you through the day. People usually recharge their devices overnight, so a battery life of twelve hours tends to be enough.

- *Memory.* Most devices have adequate memory for their basic functions (e.g., calendar, phone book), but if you're working with large files, you may need an extra memory card option.

- *Touch screens.* Many PDAs allow you to directly input data into the screen, using a stylus that comes with the device or your fingers. Even die-hard Digital Immigrants can learn the stylus shorthand quickly. Some newer phones on the market are completely touch-screen and don't require a stylus; you can use the onscreen keyboard and let your fingers do the scrolling.

- *Other options.* Many hand-held devices offer digital cameras, word processing, and spreadsheet programs as well as compact flash cards and MP3 players for music and videos. Wireless capability allows you to connect directly with other devices and the Internet. Some gadgets have full-size portable, foldable keyboards for easy data entry on the run, and many allow the user to watch television right on the little screen.

ENTERING THE BLOGOSPHERE

Blogs, Internet logs or journals, can be updated with software that requires minimal technical background. Responses to blog postings have evolved many of these sites into virtual communities. Political blogs have shaped election results, and many businesses are revising their policies as a result of customer complaints voiced in the blogosphere.

Bloggers are often people with a passion about a topic and a message to communicate. Many people start blogging through social networking sites like MySpace and Facebook. Others begin through sites like Blogger.com or Wordpress.com, which offer a platform, domain, and hosting for free. More advanced bloggers might use a stand-alone platform that is hosted on their own domain or URL. This approach offers more flexibility and design control, but it requires greater expertise and cost. If you're a blogger, consider the following to increase the readership of your blog:

- *Market your blog.* In addition to social networking sites, you can use forums, chat rooms, paid advertisements, and search engines to drive viewers to your blog.

- *Stick to your topic.* Most visitors will view your blog because of a shared interest in the topic. If your blog is FlyFishingFor Righthanders.com, don't let your entries drift off into long discussions about your new hybrid auto purchase. That may just annoy your readers, who will be less inclined to return to your site.

- *Current events count.* Keep up with the latest news on your topic. Visitors sharing your interest will likely be reading newspapers and other blogs, so the chances that they will revisit your blog will increase if you keep up with the latest news about your topic. Be sure to research and substantiate your facts.

- *Keep a schedule.* You don't need to add an entry to your blog every day, but decide on how often you can update it, and stick to a regular schedule. Loyal readers will learn to return at those intervals when they know there will be a new posting. Of course, more frequent entries will increase the number of hits, or visits.

- *Headlines matter.* Titles are the first thing to draw in a reader and will show up on keyword searches. Keep your blog name and entry titles short and to the point.

- *Write clearly.* Simple, direct language with a clear message will attract readers. Avoid the frequent use of abbreviations or jargon.

- *Proofread your entries.* Unless you have an editor to double-check your posts, try to catch spelling mistakes and typographical errors.

INTERNET PHONING AND VIDEO CONFERENCING

Thanks to broadband Internet connections, we now have a network that makes phoning and video conferencing relatively easy. VOIP (voice over Internet protocol) services like Skype or Vonage allow you to make international phone calls at relatively low cost, helping family members and friends stay in touch when they are separated by long distances. These technologies have even fostered a new form of Internet dating, whereby potential couples separated by long distances get to know each by Internet phone before taking the plunge for a plane ticket and a first face-to-face meeting.

People have avoided using Internet phone services not only because the voice quality can be suboptimal but because service is dependent on the Internet connection. If you do opt for this type of service, you may want to have at least one standard telephone land line for emergency backup. A new form of VOIP technology called Ooma involves an initial purchase of a piece of hardware for about $400 with no additional monthly charges.

Video conferencing combines audio and video telecommunications so that two or more individuals separated geographically can hold meetings. A video camera, microphone, and speakers, as well as a telephone or Internet hookup, are needed at each site. If you plan to use video conferencing for business, keep in mind that conference attendees often have tight schedules, so plan and distribute an agenda in advance and designate one main speaker or moderator at each location.

Some sites now allow people to have face-to-face chats within a Web browser without having to bother downloading additional software. They don't even have to be online at the same time because they can leave video messages for one another.

DIGITAL ENTERTAINMENT: SWAPPING HI-FI FOR WI-FI

Digital technology and the Internet are becoming the new conduits for all forms of entertainment, diversion, and amusement. We can watch television shows and movies on demand and download music to our hand-helds in seconds. The stand-alone television console is becoming obsolete furniture, soon to be replaced by a computer-based, all-in-one entertainment/business/work-and-play center.

Social networking sites and blogs are promoting TV channels that have found a niche on the Web. Because of the costs and competition in cable television, many newer cable channels that are struggling to attract a large enough audience and sponsors have instead found a home on broadband. For example, the healthy living Lime Channel left cable in 2006 when it shifted its programming to online and video-on-demand offerings. The Black Family Channel now thrives on the Internet after struggling in the cable realm. According to research from In-Stat, the Internet TV market reaches approximately 16 million households worldwide and by 2010 is expected to grow to 130 million.

Digital Immigrants may recall the first video tape recorders, or

VCRs, which were large, heavy components that required loading a blank tape for each recording. Now we can TiVo a movie while we're at work, download the digital recording to a blank DVD, and take it with us to watch on our laptop while flying on an airplane or riding a train.

Coping with a forty-two-button remote control can be daunting, however, and many people get professional technical help to set up their entertainment systems. The consumer technology trade group, Consumer Electronics Association, has a website (Ceaconnectionsguide.com) that serves as an interactive guide to help consumers hook up a variety of entertainment system components, both digital and analog.

Desktops, laptops, hand-held devices, and smart phones are also being used to watch video clips on popular Internet hosting sites like YouTube and Google Video. Sites like Joost.com and Hulu.com bring television to the Internet by offering music videos, documentaries, and TV shows. Television networks and cable channels also offer selected shows online. No need to fret if you forgot to set your TiVo for last night's episode of your favorite show; just log on to that channel's website and enjoy your missed episode at any time of day—provided you have a broadband connection, of course.

As we gain greater control of our social and entertainment worlds through TiVo, YouTube, and advanced remote control devices, the brain's anterior cingulate and prefrontal cortex allow us to enjoy the empowerment of choosing among so many options. Neuroscientists have found that this perception of control is associated with subcortical dopamine neural networks deep within the brain. Many people admit a sense of dependence—controlled by dopamine pathways—on their nightly use of TiVos, remote controls, and other devices.

ONLINE SAFETY AND PRIVACY

The threat of computer viruses, hackers, and other online mischief has made safety and privacy important issues for technology users. To understand how our brains respond when we face a fear or threat of danger, neuroscientists have pinpointed the brain's strategic region that is affected the most: the prefrontal cortex. As our safety becomes more immediately threatened, brain activity shifts to the periaqueductal gray area, a region in the middle of the brain that controls basic survival responses of fight or flight.

Most of us know that opening an attachment in an email from a person we don't know could infect our computer with a virus. But it doesn't keep us from checking our email, nor does it raise our anxiety levels. However, if we actually receive a classically threatening email, such as "Dear Friend, Attached is the information you asked for" (*never* open one of those), it does raise our anxiety that a virus sender has singled us out.

Despite a variety of ways to increase the security of routers, networks, and computers, too many of us are still not adequately protected. Some Internet users have never installed any antivirus protection or have failed to renew their subscriptions.

Your router, which connects your computer to an Internet network, has a password for access, and that password should be changed at regular intervals. Your computer also has a firewall or content filter, which should be kept on. Many people also use passwords to protect their wireless networks.

It's a good idea to routinely run virus-checking software and to back up important files on a regular schedule. You can download the files to an external drive, an online storage site, or a CD or DVD. A quick and inexpensive way to back up a file is to save it in cyberspace by emailing it to yourself. Just remember that if it's a sensitive document, it is less secure once it has been placed in cyberspace.

Businesses are becoming increasingly concerned about data privacy. In the spring of 2006, the U.S. Department of Veterans Affairs reported that the personal data of twenty-six million veterans had been stolen after a VA employee took a computer home. Some states have enacted laws requiring companies to notify customers if an employee loses a laptop with sensitive information.

Passwords are useful, but they can also be a nuisance—especially if you forget them. Automated PC hackers trying to crack a password can test up to ten million per second, so if your password contains only eight lowercase letters, it would be compromised within six hours. When you mix in some uppercase letters, numbers, and perhaps a few additional characters, it might take the same hacker program thousands of years to discover your password.

The best passwords are secure yet memorable *to you*. Avoid passwords like your birth date, your Social Security number, or your mother's maiden name. Good passwords are long enough and senseless enough to be hard to hack or randomly guess. You may prefer passphrases, which use the first letter of each word in a phrase that has meaning only to you. For example, "My dog is a black Labrador" becomes MDI-ABL. By alternating upper-case and lower-case characters and inserting an exclamation point at the end, I now have an excellent password: *MDiabL!*.

To further heighten your computer security, pay attention to cookie files—the bits of information that an Internet site sends to your browser when you access that site. Cookies allow websites you have visited to assign unique identifiers to your computer, allowing them to send whatever they like directly to your computer. You can set your computer to routinely delete cookie files or to notify you when a cookie is being written. You may also wish to get a cookie managing program like Cookie Crusher or Cookie Cruncher.

Pornography is readily accessible on the Web, and many parents express concern about their children being inadvertently exposed to lewd images and text. The 1998 Child Online Protection Act required commercial sites to verify that users were eighteen or older through credit card registration; however, an injunction has prevented the law's enforcement. In upholding the injunction, the U.S. Supreme Court argued that software filters can effectively prevent minors from viewing pornography.

Many parents use such filters to scan and block Internet pages with words like "breast" that appear without words like "feeding" or "chicken." However, filter programs need to be configured, and many parents instead choose the most restrictive default modes, which are often too restrictive because they block sites that might help teenagers with their homework.

Perhaps one of the most effective ways of dealing with the problem is for parents to initiate discussions about the impact of online pornography and how to handle it. Teenagers bristle and tend to rebel when parents make something forbidden instead of discussing it. For younger kids who visit only a few sites likely Disney or Sesame Street, a cyber-nanny can effectively block out all the smut, and the parents can save their breath for future conversations when adolescence kicks in.

In the summer of 2007, another new twist to online privacy emerged when Google introduced street-level views for its maps in order to help users scout places they planned to visit. Unfortunately, some of the candid photos captured recognizable people getting traffic tickets, sunbathing in bikinis, getting their mail while wearing pajamas, and in other normally private moments. Google responded to complaints about the embarrassing images, but by the time the images were removed, they had already been circulated on the Web.

Many companies are now keeping an eye on their employees' online habits in order to follow government regulations designed to protect intellectual property and trade secrets. The bottom line is that your boss may be snooping on your online activities right now, so here are a few strategies to help keep you out of trouble while on the job:

- *Know the rules.* Read the fine print, and make sure you are aware of your company's policies on email and Internet use. Find out how long electronic records are saved. There may be restrictions on the use of camera phones, text messaging, or downloading software.

- *Limit surfing.* Although a certain amount of Web surfing may be part of your job, most companies view Internet surfing as a distraction from work duties.

- *Create PDFs of your sensitive documents.* Whether it's a résumé, billing statement, or letter of reference, protect your attached sensitive documents by converting them from Word to PDF files. That way, others will have a harder time altering the content should they wish to misrepresent you. If you do send a Word file, disable "track changes" so others don't view earlier versions that may contain potentially embarrassing edits.

- *Double-check posts.* When posting an online profile or blog, keep in mind that it is a public record and could end up in court. Watch what you post on social network sites such as MySpace and Facebook, because those profiles can be viewed by potential employers, college admissions officers, and co-workers.

- *Joke off line.* Think twice before forwarding a controversial or distasteful joke—it can come back to haunt you if it gets into the wrong hands. Sexual humor may be interpreted as harassment, and because there's an electronic record of it, there's no denying who sent it.

Although Internet chat rooms can be a convenient way to exchange information and learn about specific areas of interest, they also pose safety issues. Don't use your real name, and never give out personal information. If an online conversation is making you uncomfortable, don't feel the need to respond—that will only encourage the other person to persist. Simply ignore it and log off.

Many single people use Internet technology to jump-start their romantic lives. The Web can increase the likelihood of meeting Ms. or Mr. Right faster than waiting for a chance encounter at a party or an occasional blind date. However, Internet dating is no guarantee of romantic success—your date may not show up for a meeting or may mislead you about who he or she really is. Some online dating sites may attract people with questionable motives, and predators have taken advantage of young teens and other innocents who visit these sites.

CYBER MEDICINE

Advances in medical technology have lengthened how long and how well we live, and the World Wide Web may allow us to reap some of the benefits of new technologies from around the world. For example, the U.S. Food and Drug Administration (FDA) is considering approval of a surgically implanted device that measures pressure inside the heart along with heart rate and body temperature, and then wirelessly transmit the data over the Internet to the doctor. This particular device would allow physicians to monitor patients suffering from unstable heart disease without having to admit them to the hospital.

Searching for Health Information

The Internet has already become an important source of health information for many of us. Going online lets us keep up with the latest medical breakthroughs, find health care referrals, stay in touch with our doctors without having to wait on the phone or schedule a visit, and even purchase medications.

The Pew Internet Project reported that eight out of ten American Internet users have searched for health information online, and more than one-third of them said that the information helped determine whether or not they saw a doctor. Despite the frequent online searches for medical information, as many as eighty-five million Americans gathering online health advice generally don't bother to check the reliability of the information, and sometimes the health information posted online can be misleading, outdated, or plain wrong.

To increase the likelihood of obtaining valid answers to health questions, don't just stick to general search engines like Google and Yahoo. Instead, look to health consumer sites like WebMD or those from reputable nonprofit organizations (see Sidebar) that have a system of quality control. If a Web address indicates a government (.gov), educational (.edu), or nonprofit (.org) site, the chances are better that the information is reliable than if it were generated from a for-profit (.com) site.

SOME RELIABLE ONLINE HEALTH SITES

AARP.org/health—supported by the national advocacy group for older Americans, this site provides information on medical insurance, medications, and healthy lifestyles.

Americanheart.org—visitors can learn about the treatment and prevention of heart disease on this site, sponsored by the American Heart Association.

Cancer.org—this American Cancer Society site offers information on various forms of cancer, support groups, and clinical trials.

Cdc.gov—this official site of the Centers for Disease Control and Prevention focuses on communicable diseases and public health.

Clinicaltrials.gov—supported by the National Institutes of Health, this site provides information on both government-sponsored and privately sponsored clinical trials.

Diabetes.org—supported by the American Diabetes Association, this site highlights research and health information related to diabetes.

Familydoctor.org—the American Academy of Family Physicians sponsors this site, which answers general health questions.

Healthfinder.gov—sponsored by the U.S. Department of Health and Human Services, this site's search engine can help find the answer to a multitude of health questions.

Mayoclinic.com—useful health information from the Mayo Clinic.

Medlineplus.gov—the National Library of Medicine provides information on a range of health topics.

WebMD.com—this well-edited, for-profit site posts reliable health information.

www.aging.ucla.edu—the site for the UCLA Center on Aging, which posts newsletters and links on healthy aging topics.

To ensure that the information you get online is up to date, look for a notation indicating the date of the posting. Also, be sure to double-check content from multiple sources. Organizations such as the Swiss nonprofit organization HON (Health on the Net—www.hon.ch) or URAC (Utilization Review Accreditation Commission—www.urac.org) rate and set standards for online medical information.

Some people, often hypochondriacs, imagine that every little tension headache heralds an oncoming brain tumor. They tend to over-interpret minor symptoms and to become hysterical over physical complaints that others generally ignore. Brain imaging studies of hypochondriacs show that areas deep within the brain—the caudate nuclei and the putamen—are smaller and less active. These regions also control other forms of anxiety, particularly from obsessive-compulsive disorder, wherein people ruminate unnecessarily about trivial and irrelevant details.

People with a tendency to focus on their physical symptoms—whether or not they are full-blown hypochondriacs—find that their Internet searches only fuel their fears. If you Google "back pain," you'll get over a hundred million entries. Focusing your search on "low back pain" will narrow your results down to only thirty-two million websites offering information. Before long, not only will you be playing doctor, but you'll develop one of the medical profession's own common ailments, "medical studentitis," the hypochondriasis that afflicts many young doctors in training when they start learning about the various possible illnesses that their everyday symptoms could suggest. The best cure for this ailment is to turn off your health search browser and discuss your actual symptoms and concerns with a doctor you trust.

Finding Doctor Right

Web-based resources are now available to help us find doctors or health care systems. It's still a good idea to ask a friend, neighbor, or family member for a recommendation, but now we can also use free online tools and some fee-based services to guide us to high-quality medical care.

Health care plans like Aetna and UnitedHealth Group are collecting data that rank physicians in their systems based on their quality of care records. You might check the website of your health care provider or insurance company to see if a rating system is in place. You can also look up a particular doctor's licensing status on the Federation of State Medical Boards site (www.fsmb.org/directory_smb.html). Most state medical boards provide online search tools that tell you whether a physician has been disciplined and why. For a small fee, you can search all records in the United States at the same time through Docinfo (www.docinfo.org). If you're looking for a physician who is a specialist, you can search the American Board of Medical Specialties (www.abms.org/newsearch.asp) for more information about their specialty status. Another resource that charges an annual fee is Consumers' Checkbook (www.checkbook.org), which locates doctors recommended by other doctors.

Hospital Compare (www.hospitalcompare.hhs.gov) can help you find highly ranked hospitals in your area, based on several quality measures collected by Medicare. Leapfrog Group (www.leapfroggroup.org) also provides hospital ratings.

Connecting with Your Doctor Online

Although surveys show that most patients express interest in communicating with their physicians online, many doctors remain concerned about privacy issues and about potential medical-legal risks. Lack of insurance reimbursement for email time also dissuades physicians from using email to communicate with patients.

Despite these issues, the speed and efficiency of email can improve many forms of doctor–patient communication. Email creates a written record of the communication and eliminates any doubt about what was advised or prescribed during the interaction. Medication names, specific instructions, and advice are automatically archived and readily retrievable. Frequently used educational handouts can be attached or relevant Web links inserted into the email. Given the gradual increase in this form of doctor–patient communication, organizations like the American Medical Association have adopted guidelines to help physicians use email with their patients in an efficient and safe way.

If you and your doctor do use email to communicate, consider the following tips for more effective exchanges:

- Discuss in advance the types of correspondence (e.g., appointment scheduling, prescription refills, medical information) and topics (e.g., mental health issues, HIV) that are appropriate for email.

- When you initiate an email query, clearly indicate the topic in the message subject line (e.g., billing question, appointment, medical advice).

- Include your name and patient identification number in the body of the message.

- Keep your messages brief and to the point (see previous section on electronic mail).

- Maintain an electronic or paper copy of all email communications between your doctor and yourself for future reference.

- Remember the limitations of email, including the inability to communicate subtle emotional aspects of an illness or condition. When necessary, don't hesitate to call your physician or go see your doctor in person.

Online Drugs

Although it's easy to purchase medicines online, it is important to take precautions to ensure that the transactions are safe. Try a Yahoo search of the old anti-anxiety drug Valium and you'll find over eleven million entries, including drug-peddling websites promising "OFF-SHORE PHARMACY WITH NO QUESTIONS!" Many of these online medicine vendors are not licensed pharmacies in any of the United States. Others are not pharmacies at all and make no effort to protect your personal information. Although legitimate Internet pharmacies do exist, many other sites sell dangerous and nonapproved products. Ordering drugs without a physical examination puts you at risk for misdiagnosis and side effects from taking unnecessary medicines, but many online pharmacies require only that you complete a brief Web questionnaire before you purchase a medication.

Often online drug peddlers provide no street address or phone number for their companies and give you no way to know the exact ingredients of the drugs, where they are manufactured, and what sort of quality control is used in their production. Some of these sites sell counterfeit drugs with dangerous additives, or drugs past their expiration date.

Despite these potential dangers, it is possible to safely purchase medicines online. Before making such a purchase, it's a good idea to check out a useful website supported by the FDA Center for Drug Evaluation and Research (www.fda.gov/buyonlineguide), which offers consumer safety guidelines for buying prescription medicines online. Here are some of these guidelines:

- Always meet with your doctor to discuss the benefits and potential side effects of any medicine he or she prescribes.

- Have a physical examination before you take a medicine for the first time.

- Make sure that the prescription website is a state-licensed pharmacy in good standing, and located in the United States. You can find a list of state boards of pharmacy on the National Association of Boards of Pharmacy website (www.nabp.info).

- A safe website should have a licensed pharmacist available for answering questions. It should also require a prescription from a licensed physician, as well as provide its street address and telephone number.

- Check for privacy and security policies that are easy to find and understand.

- Be sure to find out whether your medical insurance company has its own online pharmacy.

BRAIN STIMULATION: AEROBICIZE YOUR MIND

Thanks to recent research suggesting that daily mental stimulation may improve memory and brain health, middle-aged and older individuals are venturing beyond their daily crossword and Sudoku puzzles to new technologies for brain stimulation and neuron protection. The virtual online interactive games that kids and young people play produce both productive and negative mental effects (Chapter 3), but now we're seeing a new genre of electronic games targeted toward boomers and elders, based on the positive results of studies performed on laboratory animals and human volunteers. These studies suggest that working out our neurons through mental puzzles and games can strengthen them and may even stave off age-related diseases such as Alzheimer's. Although some of the scientific evidence is circumstantial and does not prove a cause-and-effect relationship, the mental stimulation is fun and enjoyable, and the risks are minimal—provided that the individual does not become addicted to the electronic games.

Brain scanning studies using PET and functional MRI imaging show a pattern of increased brain efficiency when volunteers train their brains with mental aerobics or memory training techniques. Our UCLA research team found that an area of the frontal cortex, which controls very short-term memory, has significantly greater efficiency after memory training. Other neuroscientists have found that brain games show similar patterns in the gray matter rim throughout the brain. Brain fitness is very much like physical fitness. The more you work out, the more you can do, and the less exertion it takes for you to do it.

When searching for fun and challenging mental aerobics, look for activities or technologies that are at the right difficulty level for you. If the game is too easy, it will not stimulate your neuronal connections, and you'll get bored. If it is too challenging, you'll become frustrated. Rather than strengthen your synapses, your body will secrete stress hormones that can lead to mental fatigue.

There are many games to choose from. In addition to some websites listed in Appendix 3, the following are a few of the options currently available:

- *Brain Games (Radica/Mattel; $20)* is a basic, inexpensive hand-held electronic game that I helped develop. It allows you to cross-train your brain by switching from right-brain (visual-spatial) to left-brain (verbal) exercises. You can also set the difficulty at six different levels. Like many of the other brain stimulation devices, it allows you to track your progress. In addition, Brain Games teaches my basic memory technique (Look, Snap, Connect) for everyday memory challenges. Brain Games was selected for inclusion in the 2007 Reader's Digest "America's 100 Best" in the Best Brain Teasers category.

- *Brain Age (Nintendo; $150)* was originally developed in Japan and imported to the United States. It requires the Nintendo DS hand-held game system and includes well-known puzzles like crosswords and Sudoku, as well as other brain training exercises. Charts and graphs allow the player to gauge progress and lower his or her "brain age," although the validity of the player's actual brain age is questionable. It has a speech recognition component, which is a nice touch; but unfortunately, many reviewers note that it doesn't work consistently.

- *Brain Fitness Program (Posit Science; $400–$600)* offers cartoonlike scenarios with family, travel, and other themes geared for seniors. The company has tested the game in a randomized trial of older adults and found some improvement in memory and attention tests in research volunteers.

- *[m]Power (Dakim, Inc.; home version—$2,500)* is a cognitive exercise system originally developed for assisted living facilities but now available in a home version. It uses a touch screen, incorporates technology that automatically adjusts difficulty levels, and is connected to the company's content providers via the Internet so programming material is continually updated. [m]Power combines original content with memory-invoking images, movies, music, and sounds that provide mental stimulation and brain training. It is currently being tested in clinical studies for effectiveness in improving memory and mood.

Recent research suggests that stimulating our brains with these kinds of games, puzzles, and exercises can improve cognitive abilities as well as the efficiency of the neural circuitry controlling them. In fact, any of the technology skills you practice in this toolkit will stimulate your neural pathways. The key is to vary the stimulation to strengthen your neural networks. In addition to these technological brain stimulators, many types of low-tech activities will also stimulate the mind and help build stronger synapses, whether it's playing chess, traveling to foreign lands you've never visited, taking up the tuba, or volunteering at your local hospital.

Nine

BRIDGING THE BRAIN GAP:

Technology and the Future Brain

> *It has become appallingly obvious that our*
> *technology has exceeded our humanity.*
> Albert Einstein

At this critical moment in brain evolution, Digital Immigrants and Digital Natives need to share one another's knowledge and experience to move forward and thrive. As new technology becomes a seamless component of nearly every aspect of our lives, it's becoming essential to interface both high-tech know-how and personal interaction skills in order to enhance job efficiency, while still maintaining our humanity. This is not only possible but necessary.

UNDERSTANDING THE GAP

The face-to-face communication exercises and the technology toolkit in the previous chapters are important steps toward bridging the brain gap. However, younger and older generations differ not only in their technology and social savvy but also in their values, expectations, aspirations, and personal experiences. To better understand these influences, we should consider that within the category of Digital Natives there are two subgroups, the Millennials and the Generation Xers, while Digital Immigrants comprise two other subgroups, Baby Boomers and Seniors.

The nearly eighty million Millennials (AKA Generation Y), born between 1981 and 2000—now mostly in their teens and twenties—appear to be the most technologically sophisticated. They value financial success as well as balancing work and play. The fifty million Generation Xers, born between 1965 and 1980 (currently in their late twenties to early forties), are often characterized as self-reliant and willing to take risks. If they don't like a particular job, they'll chuck it, just as they would with an outdated piece of technology. The eighty million Baby Boomers, born between 1946 and 1964 (currently mid-forties to early sixties), grew up in relative prosperity but also during the upheaval of the 1960s, so they are willing to challenge authority. Boomers work long hours to achieve what they want, and they value their individualism. Today's Seniors, born before 1946, are the most traditional group and the most likely to stay with the same job over the years. They have great respect for authority and tend to be the least technologically sophisticated.

Journalist Carol Hymowitz has described how businesses can find ways to narrow the culture gaps among these generations by varying management styles according to the expectations and values of each particular subgroup. For example, Generation Xers learn new information best through familiar Web-based formats, while Boomers prefer instruction in traditional classroom settings. A Generation X office manager might address a Senior employee with deference and formality, while motivating a Boomer colleague with praise about the long hours she worked on her last project. Pairing a technology-skilled Millennial with a corporate-savvy Boomer on a project may help the Millennial improve interpersonal communication skills, while the Boomer picks up some computer programming abilities.

Joe C. had never been very good with gadgets. His forte was understanding and motivating people, which had gotten him to such a high position in his company. He shunned any new technology until he was simply forced to learn it. His wife operated the TiVo at home—he'd rather read a book than waste time trying to figure out another darn contraption. His secretary answered any email that came for him, and the closest he got to a computer screen himself was to view an occasional YouTube video that his grandson would insist on showing him on weekends.

When Joe's secretary unexpectedly retired for health reasons, he had trouble finding a replacement who knew how to handle his correspondence—including email—the way he liked it done. Exasperated after running through half a dozen replacements (after all, how hard is it to write a decent memo?!), he decided to respond to the letters and emails himself. After accidentally deleting half of the emails in his inbox, Joe felt overwhelmed. He called down to the mailroom and asked for the new kid, Bobby, to come up.

"What's up, Mr. C?" Bobby asked. Joe replied, "I haven't got a clue how to use this stupid email. What ever happened to the days when secretaries took dictation and wrote shorthand, and people mailed letters?"

Bobby wasn't sure . . . he had heard of shorthand, but what was "dictation"? In any event, Bobby, a true computer geek, offered to help. He patiently showed Joe the basics of using Outlook email. Joe was more motivated (actually, more desperate), now that he couldn't find a secretary, to learn to write his emails himself. Joe quickly learned about attachments, message threads, and managing his inbox.

Bobby gave Joe a daily tutorial for the next week, and he threw in tips on search engines as well. In appreciation, Joe took Bobby under his wing and began mentoring him on business politics and negotiation. He invited Bobby to a few lunches with clients and colleagues, and Bobby was quick to pick up some business savvy and client management skills. Watching Joe in action, Bobby learned to appreciate the subtleties of face-to-face business interactions, and he developed a new respect for his boss's skills and experience.

As is typical of many Digital Immigrants, Joe finally developed some technology proficiency when he had little choice in the matter. Many Digital Immigrants are happy with the way they do business and will find ways to avoid learning new technology skills. But when forced to make the extra effort, they tend to learn rapidly. Having an accomplished Digital Native show the way not only was effective in jump-starting Joe's email abilities but also offered Bobby a unique opportunity to improve his social skills, as well as his face-to-face business abilities, through Joe's mentoring.

Such innovative approaches in the workplace can be equally effective in our personal lives, and many families naturally draw on the unique skill sets of multiple generations. For example, Digital Native Millennials commonly help their grandparents get started on a first computer or cell phone, while parents tend to mentor their young children and teenagers in basic social skills.

SOCIAL SKILLS UPGRADE FOR DIGITAL IMMIGRANTS

Finding creative ways to bring together younger and older individuals will help optimize the neural circuitry for both generations. Tech-savvy and emotionally intelligent minds can also complement one another's abilities within a generation. In recent years, many Digital Immigrants have become so immersed in new technology that they have lost some of their ability to connect socially. They suffer from the same symptoms as typical technologically overloaded Digital Natives, who often feel isolated from others except when online.

Steve was finally able to get home at a decent hour—eight o'clock. He just wanted to check his email one more time before dinner. As he darted off to his study he noticed that the house seemed extra-quiet. Maybe his wife Shirley had taken the kids out and he had missed dinner again. It certainly wouldn't be the first time he'd done that. He could grab some leftovers when he had a chance. A half-hour later, Steve finally pulled himself away from his computer to look for a snack and noticed a note on the kitchen table from Shirley. She curtly informed Steve that she had taken the kids out to dinner and that if he didn't start leaving his laptop at the office, she and the kids would be leaving the house for good.

Terrified that he was about to lose his family, Steve agreed to an initial couples therapy session, wherein he and Shirley each pleaded their case—Shirley never appreciated how hard he worked for the family. Steve had no clue that ever since they'd gotten DSL high-speed on their home computers, he had become emotionally absent from their marriage. Shirley couldn't take it any longer. The loud buzz of Steve's BlackBerry interrupted the session four times. Shirley's exasperation was palpable.

After a few sessions, the therapist learned that Steve had devoted his career to building an ad agency that had taken off when he focused on Internet marketing during the dot-com boom. His ego thrived on the financial and personal success at work, and the more time he spent working, the more success he experienced. Thanks to his innate ability to use new technology, he was able to stay connected with clients and employees day and night.

As his business excelled, however, he become less interested in playing with his kids and admittedly had been drifting emotionally away from his wife. She had long ago given up her career as a high school English teacher to rear their two boys, who would both be leaving for college in the next two years. With an empty nest just around the corner, Shirley found herself questioning whether she wanted to stay in a relationship that kept her yearning for true companionship. She and Steve no longer talked about feelings—or anything substantive, for that matter.

Steve complained that whenever he tried to talk with Shirley, she merely criticized him—he worked too much, or he didn't understand what she was going through, or how she felt alone in rearing their boys, and so on. He admitted that sometimes he buried himself further in his work to avoid the frustration he felt when they tried to connect.

The therapist attempted to give them a better perspective on what they were going through and suggested that they continue couples therapy weekly to help them communicate better. He gave Shirley and Steve some homework to do between sessions, including face-to-face communication exercises. At first, the exercises led to arguments and hurt feelings. But with time, they got better at it. Shirley found that assertiveness role-playing helped her talk more directly about her feelings without being too aggressive and pushing Steve away. Steve found that nonverbal communication exercises helped him to recognize when Shirley was upset about something she was reluctant to bring up. That was a cue for him to ask directly about her feelings, which made her feel that he cared. She soon started feeling closer to him emotionally and he responded in kind.

But the turning point in bringing them closer together again came when Steve was able to admit that his work and technology habit had begun interfering with his life—it had truly become an addiction. He got help from an Internet addiction support group to curtail his Web surfing, emailing, and texting at home and to turn his BlackBerry off during family time. Shirley and Steve also signed up for a yoga class together, which helped Steve gain a sense of inner calm throughout the day. They began enjoying hanging out together again, as well as spending time with the boys as a family.

After about six months of therapy, Shirley and Steve were no longer afraid of the empty nest. In fact, they were getting excited about it. Steve no longer felt like being on the Internet all the time—he was enjoying his marriage and home life. Shirley went back to teaching, and if she did leave Steve a kitchen note, it was usually a love note, just as in the old days.

Although Digital Immigrants have the advantage of early life training in social skills and direct communication, too much exposure to new technology can create an imbalance in their professional lives and personal relationships, just as it does with many Digital Natives, who have the added complication of much less early training in interpersonal communication. Solutions will vary according to each individual, but all will involve finding a balance between adapting to new technology and nurturing our unique people skills and sensibilities.

Although current technology tends to stimulate only isolated brain regions, new scientific evidence supports its potential ability to strengthen the complex neural circuitry that controls social interaction. Studies from the University of Washington Institute for Learning and Brain Sciences show that when volunteers play an interactive computer game involving other human players, brain regions that control social skills light up. In addition, some gamers interacting in virtual worlds actually react to subtle interpersonal (yet virtual) cues like eye contact and body language. This research suggests that we may be able to develop programming to improve human contact skills that have developed over centuries of face-to-face interactions. The human brain does not have an on/off button to tell it to process things differently just because it sees a face on a computer screen rather than in person. The future brain, which may be with us sooner than decades away, will reflect the outcome of how our brains are evolving today.

THE FUTURE BRAIN

As Digital Natives and Immigrants learn to come together rather than collide, their brain neural circuitry will adapt for the better. The weakened frontal lobes of Digital Natives suffering from video game-brain will build new neural pathways from face-to-face interactions with Digital Immigrants. Technology-naïve Immigrants will improve their multitasking skills as they increase their exposure to younger Digital Natives.

As our society bridges the brain gap the future brain will emerge. Not only will this future brain be tech-savvy and ready to try new things, it will have mastered multitasking *and* paying attention and fine-tuned its verbal and nonverbal skills. It will know how to assert itself as well as express empathy, have excellent people skills, and be able to nurture its own creativity.

Today we may marvel at the extraordinary technological advances of the digital age and how the high-tech revolution has dramatically altered our culture and our brain's neural pathways. But if technological advances continue on their current trajectory, the near future may make today's developments seem trivial, if not somewhat unsophisticated. The computer keyboard and mouse may be remembered as crude tools that caused annoying wrist and finger injuries as we enter an age when the future brain directly controls email, Web searches, and computer games through mind power alone. One can imagine a Future-Brainer musing, "Remember when Google was free?" After all, directory assistance was once free, as was dialing the operator. Ah, but I date myself.

Researchers have already developed a neurochip that·links together living brain cells and silicon circuits. The electronic currents in the semiconductor material of the chip record the electrical currents of the neuron, allowing direct communication between living cells and machines.

Scientists recently trained epileptic patients to control a computer cursor with their thoughts alone. While awaiting brain surgery, the brain surfaces of these patients were fitted with small signal-detecting electrodes, and the patients were asked to control the movements of a computer cursor linked to the electrodes. Though the patients experienced initial difficulties, eventually they were all able to control the cursor on the computer screen with 70 percent accuracy by merely imagining the motor movements involved in the task. As such brain–computer interface research advances rapidly forward, it will not only help us find ways to prevent and repair neural damage but also lead to an era when our minds will directly control electronic devices—the post-keyboard age.

Brain–computer interface technologies detect and translate the brain's physiological electrical signals in order to control an output device, such as a keyboard, a computer cursor, or even a prosthetic limb. Initially developed to assist people with severe motor disabilities, these methods could lead to the next evolutionary leap in human brain development.

Researchers have used such technology to hook up a human volunteer to mentally type into a computer at up to fifteen words a minute—about half the rate of writing by hand. As silicon-based technology picks up speed in the next few years, we can anticipate neurochips that allow people to mentally write on computers at speeds approaching normal speech.

It doesn't have to involve brain surgery to get digitally hooked up. Rather than implanted neurochips, EEG electrodes can be placed on the scalp's surface to monitor and translate neural activity. German neuroscientist Niels Birbaumer has developed implant-free technology that enables people to communicate by reading brain waves through the skin. While volunteers were hooked up to this skin-reading technology, researchers used functional MRI scanning to measure their cerebral blood flow and provided the moment-to-moment results to the volunteers, who were then able to control their brain wave output and play the computer game Pong, completely hands-free.

Brain-reading technology using pulsed ultrasonic signals to transmit information directly into the mind is under development, and Pentagon scientist Stu Wolf believes that within the next few decades we'll be wearing computer headbands for "network-enabled telepathy" that will allow us to transmit our thoughts directly from our minds, through the Internet, into the mind of someone else, also wearing a headband.

In other experiments, neuroscientists are refining methods to stimulate and measure brain function. Our current strategies for detecting brain activity involve monitoring relatively large functional brain regions and neural pathways. Even using the tiniest electrodes, we still stimulate large groups of cells, charging up millions of them from just a single electric pulse.

Not far into the future, we will have the capacity to monitor and stimulate brain activity of individual cells or neurons. Scientists already have a new apparatus that uses a photosensitive protein controlled by a laser down to the millisecond, the time dimension of a brain cell's natural communication speed. This technology will permit the manipulation of individual neurons through the laser's stimulation. The cure for the senior moment of the future brain may be as simple as turning on a laser light switch. And, of course, we'll soon be checking and correcting our neural circuitry through a remote control—perhaps the same device we use to keep track of our TiVo playlist.

As our computers get faster and more efficient and cyber-brain devices become the norm, rather than struggling with a generational brain gap we may be facing the computer—human brain gap, which has been a popular theme of science fiction for years. For research or recreation, future generations, packing future brains, will likely create and play in virtual-world computer simulations of their ancestors—us.

For now, the digital technology train is speeding forward, and all of us, eventually, will be hopping on board. New technology can not only increase our efficiency but simplify our lives and actually be fun. As we anticipate and manage the pitfalls, such as high-tech addiction, video game-brain, and too much multitasking, bringing together Digital Natives and Digital Immigrants should continue as one of our most pressing priorities. As we bridge the brain gap and learn to communicate and work together at all ages, we'll be poised to adapt to whatever new advances come our way. As a result, we will not only survive the technological alteration of the modern mind but thrive because of it.

HIGH-TECH GLOSSARY

Adware—Software applications that display advertising banners and often track a user's personal information, passing it on to third parties. Adware can significantly slow a computer's performance.

Analog—Refers to technology that takes an audio or video signal and translates it into continuous electronic pulses. By contrast, digital technology breaks up the signal into a binary format, wherein the audio or video data are represented by a series of 0's and 1's. Record albums use analog technology; iPods use digital.

Antivirus software—programs that detect and eliminate computer viruses, and repair or quarantine already infected files (also see *virus*).

Application—any software that carries out automated functions (e.g., graphics, word processing, databases).

Attachment—any file, usually an image or document, that is added to an email.

Baby boomers—the generation born between 1946 and 1964, currently mid-forties to early sixties.

Backup—copies of files saved on a separate disk, separate system, or the Internet for protection against loss or damage to the original files. Backup stored off site protects against environmental file damage from floods, tornados, fires, and other natural disasters.

Bandwidth—A communication channel's capacity to transmit during a defined time period, often expressed as bits per second.

Bitmap—A type of graphic file that depicts high-resolution images.

Bitnet—see *Usenet*.

Blog—an online journal (short for "weblog") that is regularly updated, with the most recent postings appearing first. Blogging is the activity of updating a blog, and someone who keeps a blog is known as a blogger.

Bluetooth—technology that allows the user to connect and exchange information between mobile phones, laptops, and PCs over a short-range radio frequency. Bluetooth lets the user connect a wireless earpiece, microphone, or speaker to a cell phone, so the user can talk hands free while driving or walking down the street.

Broadband—any high-speed network connection, such as a cable modem or digital subscriber line (DSL).

Browser—a search engine program for finding information and sites on the Internet. Microsoft's Internet Explorer, Netscape's Navigator, or Apple's Safari are commonly used browsers (also see *search engine*).

Buddy list—a collection of friends and other screen names in an instant message program.

CD-ROM—short for compact disc, read-only memory, which will store up to 700 MB of computer data.

Central processing unit (CPU)—the CPU is essentially the computer's brains, where most programs are run.

Chat room—a virtual room where people can communicate in real time while on the Internet.

Cookies—pieces of information saved on the computer's hard disk, which a website can send to a computer browser when it accesses the site. When your browser accesses that same site in the future, the cookies facilitate communication back to that website.

Craigslist—a free, online classified advertising service.

Cyberspace—synonymous with the Internet, a term that refers to the conceptual space the new technologies have created.

Digital—data in the form of discrete symbols translated from electronic or electromagnetic signals as binary 0's and 1's (also see *analog*).

Domain—the address of a website that specifies commercial sites and their geographic locations (e.g., .com, .uk, or .ca), as well as government (.gov), educational (.edu), nonprofit (.org), and Internet-related (.net) sites.

DSL—digital subscriber line, a digital communication network that provides faster speeds than analog telephone wires.

DVD-ROM—(also known as digital versatile disc and digital video disc) popular optical disc storage media format used for data storage. Its main uses are for movies, software, and data archiving. Most DVDs are of the same dimensions as CD-ROMs but store more than six times the data.

E-commerce—short for electronic commerce or any Internet-based purchase of products or services.

Email—short for electronic mail, which can send messages anywhere in the world, along with attached images, documents, or sounds.

Encryption—a data security technique used to protect information from unauthorized inspection or alteration. A key or password is needed to read an encrypted file.

End user license agreement (EULA)—a contract between a software vendor and a user. EULAs usually appear in dialog boxes when software is first opened, and the user must check "I accept" before proceeding.

Ezine—an online magazine or newsletter.

File formats—the method of data organization within a file. A three-letter or four-letter extension that follows a period after the file name will define the type of file format, which could include any of the following:

- .dat—a raw data file with unformatted text.
- .doc—Microsoft Word file
- .exe –an executable file, which can run a Windows operating system program. Open these files with caution, because they may contain damaging computer viruses. The Macintosh counterpart is .app.
- .html—short for hypertext markup language (sometimes written as htm), which is a format designed for Web browser viewing.
- .jpeg—short for joint photographic experts group (also written as .jpg), a format that reduces the size of an image so it is more readily transportable via email.
- .pdf—short for portable document format, a file format that preserves the fonts, formatting, colors, and graphics of a document, regardless of the application used to create it. Adobe Acrobat Reader is needed to view these documents and can be downloaded from Adobe.com.
- .tiff—short for tagged image file format, which is an image format suitable for high-resolution images.

File-sharing programs—software that allows multiple users to simultaneously access the same files. Napster, iMesh, and Limewire are examples of these programs, which have been used to illegally upload and download music and video files.

Firewall—a network link that limits access among networks by inspecting transmitted data packets and completing only the transmission of authorized data.

Forums—see *news groups*.

FTP—short for file transfer protocol, a method for transferring files between computers through the Internet.

Generation X—the generation born between 1965 and 1980, currently in their late twenties to early forties.

Hacker—originally, a term of respect for computer programmers and designers, but now referring to an unauthorized user who breaks into systems, destroys data, or otherwise harms computers and networks.

Hard drive—a high-capacity digital storage device that contains a computer's recorded data on a self-contained magnetic disc.

Hit—a recorded visit to a website. The number of hits is often used as a measure of a website's popularity.

Hosting (or Web hosting)—the business of housing and providing a server, as well as maintaining files for websites.

HTML—acronym for hypertext mark-up language, which is the language used to mark up or prepare material for the Web, allowing websites and browsers to communicate and exchange documents, pictures, and sounds.

HTTP—derived from hypertext transfer protocol, a procedure used to request and transmit Web page files and their components over the Internet.

HTTPS—specifies that the HTTP is enhanced by a security system. Always make sure that there is an "S" at the end of the "HTTP" when using an online shopping page.

Instant messaging (IM)—a program or service that allows users to send and receive messages almost instantly.

Internet Protocol (IP)—written as a series of numbers separated by periods (e.g., 555.24.681.222), this is the network address for a computer or website.

Internet Service Provider (ISP)—any company providing Internet access through IP addresses.

LAN—local area network or any wired or wireless network connecting computers in the same area.

Listserv—a mailing-list computer program, which forwards emails to all others who have subscribed to the list.

Millennials—the generation born between 1981 and 2000, now mostly in their teens and twenties, also referred to as Generation Y.

Modem—a device that can convert between digital and analog signals for computer communication across telephone lines.

Monitoring software—programs that allow monitoring or tracking of websites visited or emails exchanged.

MP3 player—hardware devices and their programs that store, organize, and play digital music and video files.

Multi-user dungeon (MUD)—a computer program that accepts connections from simultaneous users over a computer network and provides them with access to a shared gaming experience.

Network—computer systems grouped together in order to share software, hardware, or information.

News groups, forums—public forums that are subject based and supported through email.

Operating system (OS)—programs that manage a computer's basic functions (e.g., security, file systems, communications). On most personal computers, this is Windows or the Macintosh OS. Unix and Linux are other operating systems often used in scientific and technical environments.

Patch—a security update that a software manufacturer will release to fix existing program bugs.

PC—short for IBM personal computer, often used to indicate that the computer is IBM compatible, but incompatible with Macintosh computers.

PC card—(also referred to as PCMCIA [Personal Computer Memory Card International Association] cards or data cards) a type of expansion card used in many laptop computers for added features, such as WWAN access.

PDA—short for personal digital assistant, such as a mobile computer or hand-held device.

Platform—the combination of a particular computer and a particular operating system.

RAM—random access memory, which runs the computer's operating systems.

Router—a hardware device that connects computer networks and directs data packets to the appropriate network.

Search engine—Web-based software for searching information on the Internet, such as Google, Yahoo, Ask, Alta Vista, Hotbot, Excite, Infoseek, and Web-Crawler.

Server—a computer that delivers information and software to other computers linked by a network.

Social networking websites—websites like MySpace and Facebook that are designed to build social networks.

Spam—unwanted and unsolicited junk email, usually intended to sell something or to obtain personal information. In extreme situations, spam can temporarily disable entire computer networks.

Spyware—software that uses an Internet connection to transmit personal information over the Internet. Users often unknowingly download spyware along with another program, so even if the downloaded program is removed, the spyware remains.

Talking word processors—writing software that provides speech feedback as the user enters words on the computer keyboard.

Text messages—short messages, limited to 160 characters, sent between mobile hand-held devices and cell phones.

Trojan horse—a seemingly benign computer program containing hidden functions that can attack a computer's hard drive.

URL—(also known as a Web address) short for uniform (or universal) resource locator, which specifies the Internet location of publicly available information.

Usenet, Bitnet—collections of news groups.

Virus—a self-replicating program that not only copies itself but harmfully modifies other computer programs and functions.

Vlogs—video blogs or episodic shows produced for the Web.

VOIP—short for voice over Internet protocol: services like Skype or Vonage for Internet phone calls.

Worm—short for the acronym "write once, read many times": a self-replicating program that differs from a virus. Whereas a virus attaches to and becomes part of another program, a worm is self-contained and propagates on its own.

WWAN—a wireless wide-area network that allows you to connect to the Internet. Available via a cellular provider, the user is required to add a hardware component (often in the form of a PC card) and pay an additional subscription fee for use.

TEXT MESSAGE SHORTCUTS AND EMOTICONS

TEXT MESSAGE SHORTCUTS

The following are some common and not so common abbreviations that will increase your text message efficiency. You'll recognize some of the abbreviations from more familiar contexts, but others are unique to the new text language. These abbreviations also are frequently used for instant messaging and less often for emailing. If you have teenagers, they probably know many of these shortcuts.

Across—ax

Activate—activ8

Address—add

All my love—aml

All the best—atb

Also known as—aka

And—&

Any—ne

Any day now—adn

Anyone—ne1

Anything—nethng, anytng

Are—r

As a matter of fact—aam

As soon as possible—asap

At—@

Attention—attn

At the moment—atm

At the weekend—atw

At your own risk—ayor

Back to—b2

Be—b

Because—cos, coz, cuz

Become—bcum

Been—bn

Been there, done that—btdt

Before—b4

Beggars can't be choosers—bcbc

Being—bn

Be right back—brb

Better—btr

Between—btwn

Birthday—bday

Boyfriend—bf

Brilliant—brill

Bye for now—b4n

By the way—btw

Call me—cm

Can—cn

Consider it done—cid

Cool—c%l

Could—c%d, cud

Couple—cupl

Create—cr8

Date—d8

Debate—db8

Demand—dm&

Dictate—dict8

Dinner—dinr

Does not—dsnt

Do not—dnt

Don't know—dk

Download—dl

Easy—ezy

Email message—emsg

End of discussion—eod

Estimated time of arrival—eta

Every—evry

Excellent—xlnt

Eye—i

Face-to-face—f2f

Fast—fst

Feel—fil

Fill in the blank—fitb

Fingers crossed—fc

For—4

Forget—4get

Forward—fwd

For your eyes only—fyeo

For your information—fyi

Frankly, I couldn't care less—ficcl

Free to talk—f2t

Friends—frens

From—frm

Funny—fune

Generate—gnr8

Get a life—gal

Girlfriend—gf

Give me—gimme

Give me a break—gmab

Going to—gonna

Good—gud, gd

Good job—gj

Good luck—gl

Good morning—gud am

Good to see you—g2cu

Got to go—g2g

Great—gr8

Great minds think alike—gmta

Happy birthday to you—hbtu

Hate—h8

Have—hv

Have a good night—hagn

Have a nice day—hand

Hello—lo

How are you?—howru

How's it going?—hig

Hugs and kisses—H&K, xoxoxo

I am—im

I couldn't care less—iccl

I don't know—idk, idunno

If—f

If you say so—iuss

I love you—ilu, iluvu

I mean it—imi

In any case—iac

In any event—iae

Information—info

In my opinion—imo

In other words—iow

Into—in2

I owe you—iou

Is—s

Just—juz

Just a minute—jam

Just a second—jas

Just for fun—j4f

Just for kicks—jfk

Just kidding—jk

Keep cool—kc

Keep in touch—kit

Kiss—x

Know how you feel—khuf

Late—l8

Later—l8r

Laugh out loud—lol

Long time no see—ltns

Lots of love—lol

Love—luv

Lunch—lch

Managing director—md

Mate—m8

Message—msg

Mind your own business—myob

Mobile—mob

More to follow—mtf

Need—ned

Never mind—nvm

Next—nxt

No comment—nc

No one—no1

No problem—np

Not—nt

Number—#

Oh, my God—OMG

Okay—k

Only for you—04u

On the other hand—otoh

Or—o, r

Over the top—ott

Parents are watching—prw

Parents over shoulder—pos

Party—prt

People—ppl

Phone—fone

Please—pls

Please call me—pcm

Please tell me more—ptmm

Private message—pm

Regards—rgds

Search the Web—stw

Second—sec

See—c

See you—cu

See you later—cul, cul8r

See you online—cyo

See you soon—cus, sys

See you tomorrow—cu2moro

See you tonight—cu2nite

Significant other—so

Sleeping—zzzz

Smile—smyl

Someone—sme1

Sooner or later—sol

Sorry—sry

Speak—spk

Stay cool—sc

Stay in touch—sit

Take care of yourself—tcoy

Talk to you later—t2ul, ttul, ttyl

Talk to you tomorrow—ttyt

Text—txt
Texting—txting
Text me back—tmb
Thanks—thx, thnx
Thank you—ty, thnq
That—dat
The—d
Then—thn
Think positive—t+
Through—thru
Time to go—t2go
To—2
Today—2day
Tomorrow—2moro
Tonight—2nite
Too much information—tmi
To tell the truth—tttt
Trust me on this—tmot
Until—til
Very—vri, v
Wait—w8
Wait for me—w84me
Waiting for you—w4u
Way too much information—wtmi

Week—wk
Weekend—wknd
Welcome back—wb
What?—?
What are you doing?—wayd
What's up?—sup, wu, wassup
When—wen
Where are you?—werru
Which—w/c
Why—y
Will—wl
With—w/, w
With all due respect—wadr
Without—w/o
With respect to—wrt
Works for me—wfm
Write back soon—wbs
Yes—y
You—u, ya
You'll be sorry—ybs
Your—yr
You're, your—ur
You too—u2

EMOTICONS

Short for "emotion(al) icon," these text-based symbols are used to represent emotions or facial expressions. To read some of the expressions, you need to mentally or physically turn the page 90 degrees clockwise. Although not recommended for business or formal communication, they have become more popular in recent years, not just among teens but among adult Internet users as well. The following are some of the more common and creative ones to be found online and in magazines and newspapers.

Annoyed	I:-(	Pain or frustration	(>_</)
Ashamed	(,_,)	Ronald Reagan	7:^]
At a loss for words	:- S	Sad face	:- (
Bored	:-O	Sarcasm	}:- }
Broken heart	</3	Silent resignation	(v_v)
Confused	%- (	Skeptical	:- /
Crying	:*- (	Smirking	;^)
Dunce-like	<:- I	Startled or shocked	=:-O
Elvis Presley	5:-)	Suspicious	(>_˜)
Evil	(_/)	Thinking about money	($_$)
Happy face	:-)	Tongue sticking out	:-P
John Lennon	//0–0\\	Very sad face	:-C
Laughing	:- D	Winking	;-)

ADDITIONAL RESOURCES

The following organizations, websites, and other resources are included to help readers build both their technology and face-to-face communication skills. Remember that "www" does not necessarily need to precede a Web address. For additional publications and resources, also check out my website, DrGarySmall.com.

HI-TECH ADDICTION

Center for Internet Addiction Recovery (netaddiction.com)—treatment center that offers a variety of services and information focusing on online addictions.

Center for Health Counseling & Resources (aplaceofhope.com/reports_internet.html)—website that offers helpful information on Internet addiction.

Sexual Addiction Support Groups (saa-recovery.org)—link to self-help organizations and twelve-step programs for people struggling with sexual compulsion, including cyber sex addiction.

A Parent's Guide to Internet Safety (www.fbi.gov/publications/pguide/pguidee.htm)—a detailed guide to protecting children from online predators and Internet pornography.

FACE-TO-FACE COMMUNICATION

AARP (aarp.org)—nonprofit, nonpartisan organization dedicated to helping older Americans achieve lives of independence, dignity, and purpose.

Administration on Aging (aoa.gov)—provides information for older Americans and their families on opportunities and services to enrich their lives and support their independence.

American Association for Geriatric Psychiatry (aagpgpa.org)—professional organization dedicated to enhancing the mental health and well-being of older adults through education and research.

American Association for Marriage and Family Counseling (aamft.org)—national nonprofit professional association for the field of marriage and family therapy.

American Psychiatric Association (psych.org)—medical specialty society that works to ensure humane care and effective treatment for all people with mental disorders.

American Psychological Association (apa.org)—scientific and professional organization that represents psychology in the United States and aims to promote health, education, and human welfare.

Dana Alliance for Brain Initiatives (dana.org)—nonprofit organization committed to advancing public awareness about the progress and benefits of brain research.

National Institute on Aging (www.nia.nih.gov)—the National Institutes of Health agency that supports research on aging and provides information about national Alzheimer's centers, and a free directory of organizations that serve older adults.

National Institute of Mental Health (nimh.nih.gov)—part of the National Institutes of Health, the principal biomedical and behavioral research agency of the United States government.

National Institute of Neurological Disorders and Stroke (ninds.nih.gov)—the National Institutes of Health agency that supports neuroscience research; focuses on rapidly translating scientific discoveries into prevention, treatment, and cures; and provides resource support and information.

UCLA Center on Aging (www.aging.ucla.edu)—university center that works to enhance and extend productive and healthy life through research and education on aging.

TECHNOLOGY TOOLKIT RESOURCES

Brain stimulation websites—brainbashers.com; braingle.com; mybrain trainer.com; mindbluff.com; neurobics.com; sharpbrains.com; syvum. com/teasers.

Cyber security (whoswatchingcharlottesville.org)—website with information and resources on Internet safety.

Educational Development Center (donjohnston.com)—information on educational software, such as Draft:Builder, for improving writing skills.

Electronic books (e.g., The Reader from Sony, Pocket eBook, Amazon's Kindle)—hand-held devices for reading and storing ebooks. Great for travelers who don't wish to lug around multiple tomes on extended vacations. Internet Archives (archive.org) and Project Gutenberg (Gutenberg.org) offer free downloads of entire books.

File attachment software (Pando.com; YouSendIt.com)—free software that dramatically increases the size of attachments you are able to send by email.

Internet phone software (Voicestick.com; Ooma.com)—Internet phone service that lets you use your computer to make calls throughout the world.

File Backup Services—online services that offer unlimited storage (mozy.com; carbonite.com).

Kurzweil Educational Systems (kurzweiledu.com)—website featuring computer programs to assist learning-challenged students in reading, writing, and studying.

Leapfrog (leapfrog.com)—toys, programs, and related educational technologies for young children and teenagers.

Scientific Learning (scilearn.com)—innovative reading and brain activation software (e.g., Fast ForWord) developed for children with developmental delays.

Screen enlarger—programs that enhance the enlargement capability of the computer screen; particularly helpful for visually impaired individuals.

SeniorNet (seniornet.com)—national organization that works to build a community of computer-using seniors.

Simple cell phones—basic mobile phones for technology challenged Digital Immigrants (e.g., Jiggerbug, Samsung A420).

Talking word processors—programs that speak aloud what is typed into the computer.

Word processor dictation—speech recognition software that translates your spoken dictation into a Word file (dictaphonedictation.com or nuance.com).

MORE HELPFUL WEBSITES

bigbook.com—the Yellow Pages online for your city of choice.

frommers.com—useful site for those who enjoy this well-known travel brand.

iea.cc—the International Ergonomics Association website, which can help you to ensure that your work station is ergonomically safe.

mapquest.com—door-to-door driving directions for most cities within the continental United States.

movielink.com—information on theatres and show times in your area so you can purchase tickets in advance.

thesaurus.com—the ultimate online synonym source site.

weather.com—flight information and up-to-date weather for your place of choice.

www.usps.gov/ncsc—a quick way to look up zip code and address information.

555–1212.com—website for finding the area code for directory assistance for a particular city.

NOTES

CHAPTER 1: YOUR BRAIN IS EVOLVING RIGHT NOW

2   *Scientists at the University of California, Berkeley:* Dong L, Block G, Mandel S. Activities contributing to total energy expenditure in the United States: Results from the NHAPS study. *International Journal of Behavioral Nutrition and Physical Activity* 2004;1:4, http://www.ijbnpa.org/content/1/1/4.
Yang C. Americans spend more energy and time watching TV than on exercise, finds new study. *UC Berkeley News.* March 24, 2004, http://www.berkeley.edu/news/media/releases/2004/03/10_amtv.shtml.

2   *Seven out of ten American homes are wired:* Gonsalves A. Number of online Americans continues to grow. *TechWeb Technology News.* May 25, 2006, www.techweb.com/wire/ebiz/188500373.

2   *A Stanford University study found that:* Nie NH, Hillygus DS. The impact of Internet use on sociability: Time-diary findings. *IT & Society* 2002;1:1–20, http://www.stanford.edu/group/siqss/itandsociety/v01i01/v01i01a01.pdf.

3   *MySpace and Facebook have exceeded a hundred million users:* Holmes E. On MySpace, millions of users make "friends" with ads. *The Wall Street Journal.* August 7, 2006.

3   *dubbed Digital Natives:* Prensky M. Digital natives, digital immigrants. *On the Horizon.* 2001;9:1–2, www.marcprensky.com/writing/Prensky - Digital Natives, Digital Immigrants - Part1.pdf

3   *literary reading has declined by 28 percent:* National Endowment for the Arts. Reading at risk: A survey of literary reading in America: Research Division Report #46. Washington, DC, June, 2004, www.nea.gov/pub/ReadingAtRisk.pdf.

4   *Professor Thomas Patterson and colleagues:* Patterson TE. Young people and news. Joan Shorenstein Center on the Press, Politics and Public Policy at the John F. Kennedy School of Government, Harvard University. July, 2007, www.ksg.harvard.edu/presspol/carnegie_knight/young_news_web.pdf.

4   *Conservation biologist Oliver Pergams:* Pergams OR, Zaradic PA. Is love of nature in the US becoming love of electronic media? 16-year downtrend in national park visits explained by watching movies, playing video games, internet use, and oil prices. *Journal of Environmental Management* 2006;80:387–93.

5   *Your brain—weighing about three pounds:* Bloom FE, Beal MF, Kupfer DJ (eds). *The Dana Guide to Brain Health.* The Dana Press. New York, 2003.

5   *Scientists have mapped the various regions: The Dana Guide to Brain Health.*

5   *number of synaptic connection sites:* Metcalfe RM. It's all in your head; The latest supercomputer is way faster than the human brain. But guess which is smarter? *Forbes* 2007;52, Volume 179, Issue 10, http://cbcl.mit.edu/news/files/forbes-article-metcalfe-5-07.pdf.

8   *accounts for the young brain's* **plasticity:** Huttenlocher P. *Neural Plasticity.* Harvard University Press, Cambridge, MA, 2002.

8   *Linguistic scientists have found:* Kuhl PK, Tsao F-M, Liu H-M. Foreign-language experience in infancy: Effects of short-term exposure and social interaction on phonetic learning. *Proceedings of the National Academy of Sciences U S A* 2003;100:9096–101.

8   *We know that normal human brain development:* Sireteanu R. Switching on the infant brain. *Science* 1999;286:59–61.

8   *Identical twins who were separated at birth:* Markon KE, Krueger RF, Bouchard TJ Jr, Gottesman II. Normal and abnormal personality traits: Evidence for genetic and environmental relationships in the Minnesota Study of Twins Reared Apart. *Journal of Personality* 2002;70:661–93.

8   *The relatively modest number of human genes:* How many genes are in the human genome? *Human Genome Project,* http://www.ornl.gov/sci/techresources/Human_Genome/faq/genenumber.shtml.

9   *One of the most influential thinkers of the nineteenth century:* Darwin C. *On the Origin of Species.* Harvard University Press, Cambridge, MA, 2001 (Chapter 4).

10 *According to anthropologist Stanley Ambrose:* Ambrose SH. Paleolithic technology and human evolution. *Science* 2001;291:1748–53.

10 *The area of the modern brain: The Dana Guide to Brain Health.*

10 *functional magnetic reasonance imaging (MRI) studies while volunteers imagine a goal:* Koechlin E, Basso G, Pietrini P, Panzer S. The role of the anterior prefrontal cortex in human cognition. *Nature* 1999;399:148–51.

11 *neuroscientists at Tokyo Denki University in Japan:* Yuasa M, Saito K, Mukawa N. Emoticons convey emotions without cognition of faces: An fMRI study, http://delivery.acm.org/10.1145/1130000/1125737/p1565-yuasa.pdf?key1 =1125737&key2=5091192911&coll=GUIDE&dl=GUIDE&CFID= 39827882&CFTOKEN=12197400.

11 *Natural selection has literally:* Ambrose SH. *Science* 2001; 1748–53.

12 *driving forces behind the Industrial Revolution:* De Vries J. The Industrial Revolution and the industrious revolution. *The Journal of Economic History* 1994;54:249–270.

12 *two American electrical engineers:* William TI, Schaal WE, Burnette AD. *A History of Invention: From Stone Axes to Silicon Chips.* Checkmark Books. New York. 2000.

13 *biologically primed to function digitally:* Levy WB, Baxter RA. Using energy efficiency to make sense out of neural information processing. *IEEE*; ISIT

2002, Lausanne, Switzerland, June 30—July 5, 2002, http://ieeexplore.ieee.org/iel5/7942/21920/01023290.pdf.

14 *A 2007 University of Texas study:* Vandewater EA, Rideout VJ, Wartella EA, Huang X, Lee JH, Shim MS. Digital childhood: Electronic media and technology use among infants, toddlers, and preschoolers. *Pediatrics* 2007;119:e1006-15, http://pediatrics.aappublications.org/cgi/content/full/119/5/e1006.

14 *A recent Kaiser Foundation study:* Roberts DF, Foehr UG, Rideout V. Generation M: Media in the lives of 8-18 year-olds. *A Kaiser Family Foundation Study.* 2005, www.kff.org/entmedia/upload/Generation-M-Media-in-the-Lives-of-8-18-Year-olds-Report.pdf.

15 *about 90 percent of young adults are frequent Internet users:* Fox S., Madden M. Generations online. *Pew Internet & American Lite Project.* December, 2005, http://www.pewinternet.org/pdfs/PIP_Generations_Memo.pdf.

18 *plunged us into a state of* **continuous partial attention:** Thompson C. Meet the life hackers. *The New York Times.* October 16, 2005.

18 *Dr. Sonia Lupien and associates at McGill University:* Pressner JC, Baldwin MW, Dedovic K, et al. Self-esteem, locus of control, hippocampal volume, and cortisol regulation in young and old adulthood. *Neuroimage* 2005;28:815-26.

19 *Under this kind of stress:* McEwen BS. Protective and damaging effects of stress mediators: Central role of the brain. *Dialogues in Clinical Neuroscience* 2006;8:367-81.

19 *Dr. Sara Mednick and colleagues at Harvard University:* Mednick SC, Nakayama K, Cantero JL. The restorative effect of naps on perceptual deterioration. *Nature Neuroscience* 2002;5:67-81.

21 *According to Professor Pam Briggs:* Sillence E, Briggs P, Harris PR, Fishwick L. How do patients evaluate and make use of online health information? *Social Science & Medicine* 2007;64:1853-62.

21 *Our UCLA research team and other scientists:* Small GW, Silverman DHS, Siddarth P, et al. Effects of a 14-day healthy longevity lifestyle program on cognition and brain function. *American Journal of Geriatric Psychiatry* 2006;14:538-45.
Haier RJ, Siegel BV, MacLachlan A, et al. Regional glucose metabolic changes after learning a complex visuospatial/motor task: A positron emission tomographic study. *Brain Research* 1992;570:134-43.

21 *Average IQ scores are steadily rising:* Flynn JR. The hidden history of IQ and special education: Can the problems be solved? *Psychology, Public Policy and Law* 2000;6:191-8.

21 *computer games can actually improve cognitive ability and multitasking skills:* Kearney P. Cognitive assessment of game-based learning. *British Journal of Educational Technology* 2007;38:529-31.

CHAPTER 2: BRAIN GAP: TECHNOLOGY DIVIDING GENERATIONS

25 *Neuroscientists at Princeton University:* McClure SM, Laibson DI, Loewenstein G,

Cohen JD. Separate neural systems value immediate and delayed monetary rewards. *Science* 2004;306:503–7.

25 *One-third of young people:* Roberts DF, Foehr UG, Rideout V. Generation M: Media in the lives of 8–18 year-olds. *A Kaiser Family Foundation Study.* 2005, www.kff.org/entmedia/upload/Generation-M-Media-in-the-Lives-of-8–18 -Year-olds-Report.pdf.

25 *Their young, developing brains are much more sensitive:* Bischof HJ. Behavioral and neuronal aspects of developmental sensitive periods. *Neuroreport* 2007;18:461–5.

25 *Young people today spend much less time reading:* National Endowment for the Arts. Reading at risk: A survey of literary reading in America: Research Division Report #46. National Endowment for the Arts. Washington, DC, June, 2004, www.nea.gov/pub/ReadingAtRisk.pdf.

26 *too much video exposure, even to these so-called educational videos:* Zimmerman FJ, Christakis DA, Meltzoff AN. Television and DVD/video viewing in children younger than 2 years. *Archives of Pediatric and Adolescent Medicine* 2007;161:473–9. Dan A. Videos as a baby brain drain. *Los Angeles Times.* August 7, 2007.

26 *Globalization and outsourcing of business:* Friedman TL. *The World is Flat: A Brief History of the Twenty-First Century.* Farrar, Straus and Giroux, New York. 2005.

27 *a baby's brain health is extremely susceptible:* Cunningham FG, McDonald PC, Norman PG (eds): *Williams Obstetrics.* 21st ed. McGraw-Hill Medical Publishing, New York. 2001

27 *Too many extracurricular activities:* Elkins D. Are we pushing our kids too hard? *Psychology Today,* 2006, http://psychologytoday.com/articles/pto–20030304–000002.html.

27 *The American Academy of Pediatrics actually recommends:* American Academy of Pediatrics. Children, adolescents, and television. *Pediatrics* 2001;107:423–6.

28 *The nineteenth-century French psychologist:* Ginsburg HP, Opper S. *Piaget's Theory of Intellectual Development.* 3rd ed. Prentice Hall, Englewood Cliffs, NJ, 1988.

28 *younger age groups are much more likely to use computers:* Microsoft Corporation. *The Market for Accessible Technology—The Wide Range of Abilities and Its Impact on Computer Use,* www.microsoft.com/enable/research/default.aspx.

29 *A recent Pew Internet survey:* Madden M. Internet penetration and impact. *Pew Internet & American Life Project.* April, 2006, www.pewinternet.org/pdfs/ PIP_Internet_Impact.pdf.

29 *Young people are more likely to use instant messaging:* Shiu E, Lenhart A. How Americans use instant messaging. *Pew Internet & American Life Project.* September, 2004, www.pewinternet.org/pdfs/PIP_Instantmessage_Report.pdf.

29 *study of more than two thousand kids and teens:* Roberts DF, Foehr UG, Rideout V. Generation M: Media in the lives of 8–18 year-olds. *A Kaiser Family Foundation Study.* 2005, www.kff.org/entmedia/upload/Generation-M-Media-in-the -Lives-of-8–18-Year-olds-Report.pdf.

30 *An estimated 20 percent of this younger generation:* Niemz K, Griffiths M, Banyard P. Prevalence of pathological Internet use among university students and correlations with self-esteem, the general health questionnaire (GHQ), and disinhibition. *Cyberpsychology & Behavior* 2005;8:562–70.

30 *In a 2006 study, Naoko Koezuka and associates:* Koezuka N, Koo M, Allison KR, et al. The relationship between sedentary activities and physical inactivity among adolescents: Results from the Canadian community health survey. *Journal of Adolescent Health* 2006;39:515–22.

30 *A recent study of children five to eleven years:* Grund A, Krause H, Siewers M, Rieckert H, Muller MJ. Is TV viewing an index of physical activity and fitness in overweight and normal weight children? *Public Health and Nutrition* 2001;4:1245–51.

30 *Dr. Robert McGivern and co-workers at San Diego State University:* McGivern RF, Andersen J, Byrd D, et al. Cognitive efficiency on a match to sample task decreases at the onset of puberty in children. *Brain and Cognition* 2002;50:73–89.

31 *Dr. Sarah-Jayne Blakemore of University College in London used functional MRI:* Blakemore S-J, Choudhury S. Development of the adolescent brain: Implications for executive function and social cognition. *Journal of Child Psychology and Psychiatry* 2006;47: 296–312.
    Den Ouden HEM, Frith U, Frith C, Blakemore S-J. Thinking about intentions. *NeuroImage* 2005;28:787–96.

32 *In 2006, a* **Los Angeles Times**/*Bloomberg poll gathered responses:* Abcarian R, Horn J. Underwhelmed by it all. *Los Angeles Times.* August 7, 2006.
    Piccalo G. Girls just want to be plugged in—to everything. *Los Angeles Times.* August 11, 2006.

33 *Patricia Tun of Brandeis University in Massachusetts found that simultaneous:* Tun PA. Fast noisy speech: Age differences in processing rapid speech with background noise. *Psychology and Aging* 1998;13:424–34.

34 *when volunteers between eighteen and forty-five years old were given a learning task:* Foerde K, Knowlton BJ, Poldrack RA. Modulation of competing memory systems by distraction. *Proceedings of the National Academy of Sciences U S A* 2006;103:11778–83.

36 *Cyber athletes compete before crowds of a hundred thousand:* Evers M. South Korea turns PC gaming into a spectator sport. *Spiegel Online International,* http://www.spiegel.de/international/spiegel/0,1518,399476,00.html.

36 *video games appear to suppress frontal lobe activity:* Mori A. *Terror of Game-Brain.* NHK Books, Tokyo, Japan. 2002. Matsuda G. Hinaki K. Neuroimage 2006; 29: 706–11

36 *more than 90 percent of all children:* AMA Report of the Council on Science and Public Health. *Emotional and Behavioral Effects, Including Addictive Potential, of Video Games,* www.ama-assn.org/ama1/pub/upload/mm/467/csaph12a07.doc.

36 *as little as ten minutes of daily violent video gaming:* Nicoll J, Kieffer KM. Violence in video games: A review of the empirical research. *American Psychological Association Annual Meeting.* August 2005.

37 *when children play video games, their brains do not use frontal lobe circuits:* Kawashima R. *Train Your Brain: 60 Days to a Better Brain.* Kumon Publishing North America, Teaneck, NJ. 2005.

38 *brain PET scan of Tetris novices:* Haier RJ, Siegel BV, MacLachlan A, et al. Regional glucose metabolic changes after learning a complex visuospatial/motor task: A positron emission tomographic study. *Brain Research* 1992;570:134–43.

38 *Dr. James Rosser and associates of Beth Israel Medical Center in New York:* Rosser JC Jr, Lynch PJ, Cuddihy L, Gentile DA, Klonsky J, Merrell R. The impact of video games on training surgeons in the 21st century. *Archives of Surgery* 2007;142:181–6.

39 *a limited amount of video:* Green CS, Bavelier D. Action video game modifies visual selective attention. *Nature* 2003;423:534–7.

40 *Learning any language in adulthood:* Szaflarski JP, Holland SK, Schmithorst VJ, Byars AW. fMRI study of language lateralization in children and adults. *Human Brain Mapping* 2006;27:202–12.

40 *functional impairments associated with aging:* Small G, Vorgan G. *The Longevity Bible.* Hyperion, New York, 2006.

41 *scored significantly higher on computer anxiety ratings:* Czaja SJ, Charness N, Fisk AD, et al. Factors predicting the use of technology: Findings from the Center for Research and Education on Aging and Technology Enhancement (CREATE). *Psychology and Aging* 2006; 21:333–52, www.pubmedcentral.nih.gov/articlerender.fcgi?tool=pubmed&pubmedid=16768579.

41 *A Pew Internet study found that 22 percent:* Fox S. Digital divisions. *Pew Internet & American Life Project.* October 5, 2005, www.pewinternet.org/pdfs/PIP_Digital_Divisions_Oct_5_2005.pdf.

41 *how the older brain differs:* Cohen GD. *The Mature Mind: The Positive Power of the Aging Brain.* Basic Books, New York, 2006.

41 *For some people, it becomes more difficult:* Cabeza R, Daselaar SM, Dolcos F, et al. Task-independent and task-specific age effects on brain activity during working memory, visual attention and episodic retrieval. *Cerebral Cortex* 2004;14:364–75.

42 *how blind people use their visual cortex:* Liu Y, Yu C, Liang M, et al. Whole brain functional connectivity in the early blind. *Brain* 2007; 130: 2085–96.

42 *when blindness comes on quickly:* Théoret H, Merabet L, Pascual-Leone A. Behavioral and neuroplastic changes in the blind: Evidence for functionally relevant cross-modal interactions. *Journal de Physiologie (Paris)* 2004;98:221–33.

43 *visual cortex can control other sensory functions:* Burton H, McLaren DG. Visual cortex activation in late-onset, Braille naive blind individuals: An fMRI study during semantic and phonological tasks with heard words. *Neuroscience Letters* 2006;392:38–42.

43 *study of nearly three thousand older adults:* Willis SL, Tennstedt SL, Marsiske M,

et al. Long-term effects of cognitive training on everyday functional outcomes in older adults. *Journal of the American Medical Association* 2006;296:2805–14.

43 **Dr. Cindy Lustig and co-workers measured brain activity:** Velanova K, Lustig C, Jacoby LL, Buckner RL. Evidence for frontally mediated controlled processing differences in older adults. *Cerebral Cortex* 2006;17:1033–46.

44 **Dr. George Bartzokis and colleagues at UCLA found:** Bartzokis G, Lu PH, Geschwind DH, et al. Apolipoprotein E genotype and age-related myelin breakdown in healthy individuals: implications for cognitive decline and dementia. *Archives of General Psychiatry* 2006;63:63–72.

44 **mentally successful older adults tend to use both hemispheres:** Cabeza R, Anderson ND, Locantore JK, McIntosh AR. Aging gracefully: compensatory brain activity in high-performing older adults. *Neuroimage* 2002;17:1394–402.

44 **Drs. Ravenna Helson and Christopher Soto:** Helson R, Soto CJ. Up and down in middle age: monotonic and nonmonotonic changes in roles, status, and personality. *Journal of Personality and Social Psychology* 2005;89:194–204.

45 **In the United States today, nearly eighty million people:** U.S. Census Bureau Resident Population Estimates of the United States by Age and Sex, http://www.census.gov/population/estimates/nation/intfile2-1.txt.

CHAPTER 3: ADDICTED TO TECHNOLOGY

48 **Internet addicts report feeling a pleasurable mood burst:** Ng BD, Wiemer-Hastings P. Addiction to the Internet and online gaming. *Cyberpsychology & Behavior* 2005;8:110–3.

48 **The brain-wiring system that controls these responses:** Di Chiara G, Bassareo V. Reward system and addiction: What dopamine does and doesn't do. *Current Opinion in Pharmacology* 2007;7:69–76.

49 **Studies of volunteers rapt in addictive video games:** Rau P-LP, Peng S-Y, Yang C-C. Time distortion for expert and novice online game players. *Cyberpsychology & Behavior* 2006;9:396–403.

49 **Previous research has shown that both eating and sexual activity:** Nestler EJ, Carlezon WA. The mesolimbic dopamine reward circuit in depression. *Biological Psychiatry* 2006;59:1151–9.

49 **the brain's executive region, known as the anterior cingulate:** Kalivas PW, Volkow ND. The neural basis of addiction: A pathology of motivation and choice. *American Journal of Psychiatry* 2005;162:1403–13.

50 **a man was fired for visiting:** Worker fired over visit to adult chat room sues IBM. *Los Angeles Times (Associated Press).* February 19, 2007.

50 **an embedded game called BrickBreaker:** Craig S, Zuckerman G. BlackBerry addicts also can't resist this little game. *The Wall Street Journal.* February 17, 2007.

50 **up to 14 percent of computer users reported neglecting school:** Associated Press. Stanford University study warns of Internet Addiction, http://abclocal.go.com/kabc/story?section=local&id=4679518.

51 *college students with difficulties adjusting:* Kanwal N, Anand AP. Internet addiction in students: A cause of concern. *Cyberpsychology & Behavior* 2003;6:653–6.

52 *The driving force of addiction:* Goldman D, Orisci G, Ducci F. The genetics of addictions: uncovering the genes. *Nature Review Genetics* 2005;6:521–32.

52 *Addiction experts have proposed criteria:* Leung L. Net-generation attributes and seductive properties of the Internet as predictors of online activities and Internet addiction. *Cyberpsychology & Behavior* 2004;7:333–47

53 *Recently, the American Medical Association:* Pham A. Gaming junkies get no diagnosis. *Los Angeles Times.* June 28, 2007.

53 *Internet addicts typically spend forty or more hours:* Beard KW. Internet addiction. A review of current assessment techniques and potential assessment questions. *Cyberpsychology & Behavior* 2005;8:7–15.

54 *Internet addiction has physical side effects:* Hakala PT, Rimpela AH, Saarni LA, Salminen JJ. Frequent computer-related activities increase the risk of neck-shoulder and low back pain in adolescents. *European Journal of Public Health* 2006;16:536–41.

54 *the rules of operant conditioning:* Beard KW. *Cyberpsychology & Behavior* 2005;8:7–15.

56 *twelve-step program designed to tackle email addiction:* Reuters. Twelve-step program aims to cure e-mail addiction, www.reuters.com/article/internet News/idUSN1943527720070220.

57 *approximately 145 million people:* Rau PP, Peng S, Yang C. Time distortion for expert and novice online game players. *Cyberpsychology & Behavior* 2006;9:396–403.

57 *While females are more likely to stay in touch:* Lenhart A. Social networking websites and teens: An overview. *Pew Internet & American Life Project.* January 3, 2007, www.pewinternet.org/pdfs/PIP_SNS_Data_Memo_Jan_2007.pdf.

57 *since online gamers tend to stick to them:* Rau et al. *Cyberpsychology & Behavior* 2006;9:396–403.

57 *In a study of the Internet heroic fantasy game Everquest:* Griffiths MD, Davies MNO, Chappell D. Demographic factors and playing variables in online computer gaming. *Cyberpsychology & Behavior* 2004;4:479–87.

57 *a fifty-three-year-old man was playing:* Alter A. Is this man cheating on his wife? *The Wall Street Journal.* August 10, 2007.

58 *4 percent of websites display sexually related material:* Spink A, Jansen BJ. *Web Search: Public Searching of the Web.* Berlin: Springer-Verlag, 2004.

58 *Forty million Americans visit:* Maltz W, Maltz L. *The Porn Trap: The Essential Guide to Overcoming Problems Caused by Pornography.* HarperCollins, New York, NY, 2008.

58 *Dr. Amanda Spink and her associates:* Spink A, Ozmutlu HC, Lorence DP. Web searching for sexual information: an exploratory study. *Information Processing and Management* 2004:40:113–123.

59   *over 70 percent of companies provide:* Dun and Bradstreet Survey. D&B study shows seven out of 10 U.S. small businesses now have Internet access. May 25, 2000, www.dnb.com/newsview/0500news8.htm.

59   *A 2006 study of more than thirty-four hundred volunteers:* Cooper A, Safir MP, Rosenmann A. Workplace worries: A preliminary look at online sexual activities at the office—emerging issues for clinicians and employers. *Cyberpsychology & Behavior* 2006;9:22-9.

59   *one out of four members of Gamblers Anonymous:* Lorenz VC, Politzer RM. Yaffee RA. Final report of the Task Force on Gambling Addiction in Maryland, www.nyu.edu/its/statistics/Docs/task_force_4.html.

59   *A 2005 survey found that 4 percent of Americans:* Ackman D. Why pick on Internet gambling? *Los Angeles Times.* October 29, 2006, www.latimes.com/news/opinion/la-op-ackman29oct29,0,656569.story?coll=la-opinion-rightrail.

60   *Congress passed the Unlawful Internet Gambling Enforcement Act:* Unlawful Internet Gambling Enforcement Act of 2006, www.techlawjournal.com/cong109/bills/house/gambling/20060929.asp.

60   *George Ladd and Nancy Petry of the University of Connecticut:* Ladd GT, Petry NM. Disordered gambling among university-based medical and dental patients: A focus on Internet gambling. *Psychology of Addictive Behaviors* 2002;16:76-9.

60   *Doctors from the Mayo Clinic:* Dodd ML, Klos KJ, Bower JH, et al. Pathological gambling caused by drugs used to treat Parkinson disease. *Archives of Neurology* 2005;62:1377-81.

61   *In China, where an estimated two million youths:* Ransom I. Chinese boot camps tackle Internet addiction. *International Herald Tribune.* March 12, 2007, www.iht.com/articles/2007/03/12/business/addicts.php.

CHAPTER 4: TECHNOLOGY AND BEHAVIOR: ADHD, INDIGO CHILDREN, AND BEYOND

64   *Digital Immigrants over forty:* Tun PA, O'Kane G, Wingfield A. Distraction by competing speech in young and older adult listeners. *Psychology and Aging* 2002;17:453-67.

64   *syndromes such as attention deficit disorder:* American Psychiatric Association. *Diagnostic and Statistical Manual for Mental Disorders* DSM-IV-TR, Fourth Edition (text revision). Washington, DC, American Psychiatric Association. 2000.

66   *An estimated 5 percent of children:* Polanczyk G, Rohde LA. Epidemiology of attention-deficit/hyperactivity disorder across the lifespan. *Current Opinion in Psychiatry* 2007 20:386-92.

66   *time that ninth- and tenth-graders spent using the Internet:* Chan PA, Rabinowitz T. A cross-sectional analysis of video games and attention deficit hyperactivity disorders symptoms in adolescents. *Annals of General Psychiatry* 2006;5:16.

66   *investigators from Kaohsiung Medical University in Taiwan:* Yen JY, Ko CH,

Yen CF, Wu HY, Yang MJ. The comorbid psychiatric symptoms of Internet addiction: Attention deficit and hyperactivity disorder (ADHD), depression, social phobia, and hostility. *Journal of Adolescent Health* 2007;41:93–8.

66 **Psychiatric investigators in South Korea:** Ha JH, Yoo HJ, Cho IH, Chin B, Shin D, Kim JH. Psychiatric comorbidity assessed in Korean children and adolescents who screen positive for Internet addiction. *Journal of Clinical Psychiatry* 2006;67:821–6.

66 **Dr. Dimitri Christakis and colleagues at the University of Washington:** Christakis DA, Zimmerman FJ, DiGuiseppe DL, McCarty CA. Early television exposure and subsequent attentional problems in children. *Pediatrics* 2004;113:707–13.

67 **National professional organizations like the American Academy of Pediatrics:** American Academy of Pediatrics. Children, adolescents, and television. *Pediatrics* 2001;107:423–6.

67 **recommendations that toddlers watch little or no TV:** Vandewater EA, Rideout VJ, Wartella EA, Huang X, Lee JH, Shim MS. Digital childhood: Electronic media and technology use among infants, toddlers, and preschoolers. *Pediatrics* 2007;119(5):e1006–15, http://pediatrics.aappublications.org/cgi/content/full/119/5/e1006.

67 **a hot new diagnosis: adult ADHD:** Spencer TJ, Biederman J, Mick E. Attention-deficit/hyperactivity disorder: diagnosis, lifespan, comorbidities, and neurobiology. *Journal of Pediatric Psychology* 2007;32:631–42.

68 **when our brains switch back:** Eppinger B, Kray J, Mecklinger A, John O. Age differences in task switching and response monitoring: evidence from ERPs. *Biological Psychiatry* 2007;75:52–67.

68 **Psychologist David Meyer and colleagues at the University of Michigan:** Rubinstein JS, Meyer DE, Evans JE. Executive control of cognitive processes in task switching. *Journal of Experimental Psychology and Human Perceptual Performance* 2001;27:763–97.

68 **Dr. Gloria Mark and associates at the University of California at Irvine:** Thompson C. Meet the life hackers. *The New York Times.* October 16, 2005.

69 **surgeons perform stressful nonsurgical tasks:** Allen K, Blascovich J. Effects of music on cardiovascular reactivity among surgeons. *Journal of the American Medical Association* 1994;272:882–4.

69 **behavior cluster has been termed the Indigo Children:** Carroll L. *Indigo Children.* Hay House, London. 1999.

70 **a related condition known as Asperger's syndrome:** McPartland J, Klin A. Asperger's syndrome. *Adolescent Medicine Clinics* 2006;17:771–88.

70 **highly intelligent children with coexisting attention deficit disorders:** Baum SM, Olenchak FR, Owen SV. Gifted students with attention deficits: Fact and/or fiction? Or, can we see the forest for the trees? *Gifted Child Quarterly* 1998;42:96–104.

71 **The exact proportion of ADHD children:** Healey D, Rucklidge JJ. An exploration into the creative abilities of children with ADHD. *Journal of Attention Disorders* 2005;8:88–95.

71 ***307 children over a seventeen-year period:*** Shaw P, Greenstein D, Lerch J, et al. Intellectual ability and cortical development in children and adolescents. *Nature* 2006;440:676.

71 ***Cornell University economist Michael Waldman:*** Whitehouse M. Mind and matter. Is an economist qualified to solve puzzle of autism? *The Wall Street Journal.* February 27, 2007.
Waldman M. Does television cause autism? www.johnson.cornell.edu/faculty/profiles/waldman/autpaper.html.

72 ***widespread brain changes that alter many aspects:*** Williams DL, Goldstein G, Minshew NJ. Neuropsychologic functioning in children with autism: Further evidence for disordered complex information-processing. *Child Neuropsychology* 2006; 12:279–98.
Koshino H, Kana RK, Keller TA, Cherkassky VL, Minshew NJ, Just MA. fMRI investigation of working memory for faces in autism: Visual coding and underconnectivity with frontal areas. *Cerebral Cortex* 2008;18:289–300.

73 ***neuroscientists at the University of Wisconsin studied an almond-shaped:*** Nacewicz BM, Dalton KM, Johnstone T, et al. Amygdala volume and nonverbal social impairment in adolescent and adult males with autism. *Archives of General Psychiatry* 2006;63:1417–28.

73 ***normal siblings of autistic children:*** Dalton KM, Nacewicz BM, Alexander AL, Davidson RJ. Gaze-fixation, brain activation, and amygdala volume in unaffected siblings of individuals with autism. *Biological Psychiatry* 2007;61:512–20.

73 ***computer games depicting violent scenes activate the amygdala:*** Mathiak K, Weber R. Toward brain correlates of natural behavior: fMRI during violent video games. *Human Brain Mapping* 2006;27:948–56.

74 ***The first outbreak was reported back in 2001:*** Mason M. Is it disease or delusion? U.S. takes on a dilemma. *The New York Times.* October 24, 2006.

75 ***defined mass hysteria as an outbreak of illness:*** Small GW, Borus JF. Outbreak of illness in a school chorus: Toxic poisoning or mass hysteria? *New England Journal of Medicine* 1983;308:632–5.

75 ***neuroimaging studies of patients with hysteria:*** Vuilleumier P. Hysterical conversion and brain function. *Progress in Brain Research* 2005;150:309–29.

76 ***In the past, newspapers, television, and radio:*** Small GW, Borus JF. The influence of newspaper reports on outbreaks of mass hysteria. *Psychiatric Quarterly* 1987;58:269–78.

76 ***the more than one hundred thousand websites:*** Alao AO, Soderberg M, Pohl EL, Alao AL. Cybersuicide: Review of the role of the internet on suicide. *Cyberpsychology and Behavior* 2006;9:489–93.

76 ***cybersuicide was brought into the limelight:*** Rajagopal S. Suicide pacts and the Internet. *British Medical Journal* 2004 4;329:1298–9.

77 ***an estimated 15 percent of the population:*** Angst J. Major depression in 1998: Are we providing optimal therapy? *Journal of Clinical Psychiatry* 1999;60 Suppl 6:5–9.

77 *social isolation clearly increases the risk:* Cacioppo JT, Hughes ME, Waite LJ, Hawkley LC, Thisted RA. Loneliness as a specific risk factor for depressive symptoms: Cross-sectional and longitudinal analyses. *Psychology and Aging* 2006;21:140–51.

77 *seniors complain about computer anxiety:* Czaja SJ, Charness N, Fisk AD, et al. Factors predicting the use of technology: Findings from the Center for Research and Education on Aging and Technology Enhancement (CREATE). *Psychology and Aging* 2006;21:333–52.

78 *control addictions and compulsions:* Lubman DI, Yücel M, Pantelis C. Addiction, a condition of compulsive behaviour? Neuroimaging and neuropsychological evidence of inhibitory dysregulation. *Addiction* 2004;99:1491–502.

78 *how young people develop their sense of self:* Allison SE, von Wahlde L, Shockley T, Gabbard GO. The development of the self in the era of the Internet and role-playing fantasy games. *American Journal of Psychiatry* 2006;163:381–5.

CHAPTER 5: HIGH-TECH CULTURE: SOCIAL, POLITICAL, AND ECONOMIC IMPACT

79 *traditional low-tech financial transactions:* Samuelson RJ. The vanishing greenback. *Newsweek.* June 25, 2007.

79 *the Web helped them make major life decisions:* Horrigan J, Rainie L. The Internet's growing role in life's major moments. *Pew Internet & American Life Project.* April 19, 2006, www.pewinternet.org/PPF/r/181/report_display.asp.

80 *In his book* **The Long Tail:** Anderson C. *The Long Tail: Why the Future of Business Is Selling Less of More.* Hyperion, New York, 2006.

80 *Dr. Christine Born and colleagues:* Born C, Meindl T, Poeppel E, et al. Brand perception: Evaluation of cortical activation using fMRI. *Annual Meeting of the Radiological Society of North America* November 28, 2006, http://rsna2006.rsna.org/rsna2006/V2006/conference/event_display.cfm?em_id=4429416..

81 *Andrew Keen describes the negative cultural impact:* Keen A. *The Cult of the Amateur: How Today's Internet is Killing our Culture.* Doubleday, New York, NY, 2007.

82 *ten most popular websites in 2007:* www.alexa.com/site/ds/top_sites?cc=US&ts_mode=country&lang=none.

82 *The Internet monitoring company Netcraft:* Netcraft.com, http://news.netcraft.com/archives/web_server_survey.html.

83 *Lee Gomes contrasted some intriguing samples:* Gomes L. What are Web surfers seeking? Well, it's just what you'd think. *The Wall Street Journal.* August 16, 2006.

83 *The top ten search terms of 2007:* Arrington M. Yahoo top searches 2007: Please, people, stop typing Britney Spears into search boxes. TechCrunch. December 3, 2007. www.techcrunch.com/2007/12/03/yahoo-top-searches-2007-please-people-stop-typing-britney-spears-into-search-boxes/. Arrington M. Google announces fastest growing search terms. TechCrunch.

December 3, 2007. www.techcrunch.com/2007/12/03/google-announces -fastest-growing-search-terms.

Mills E. Ask.com's top 10 has Google, but no Britney. C/Net News.com. December 14, 2007. Avalable online: www.news.com/8301-10784_3-9834449-7. html.

84 *study from the network information provider:* 2005 Online Holiday Shopping Update. November 27, 2005, www.comscore.com/press/release.asp?press=683.

84 *One out of every six Americans sells something online:* Lenhart L. Selling items online. *Pew Internet & American Life Project.* November, 2005, www.pewinternet .org/pdfs/PIP_SellingOnline_Nov05.pdf.

84 *According to the National Association of Realtors:* Know N. Home shoppers do their hunting online. *USA Today.* February 9, 2007.

84 *Dr. Brian Knutson and colleagues at Stanford University:* Knutson B, Rick S, Wimmer GE, Prelec D, Loewenstein G. Neural predictors of purchases. *Neuron* 2007;53:147–56.

85 *technology's positive impact on productivity:* Lohr S. Study says computers give big boosts to productivity. *The New York Times.* March 13, 2007.

87 *we are becoming a cashless society:* Samuelson RJ. The vanishing greenback. *Newsweek.* June 25, 2007.

87 *family physicians are harnessing technology:* Naik G. Faltering family MDs get technology lifeline. *The Wall Street Journal.* February 26, 2007.

87 *Bidshift Inc., a San Diego software company:* Duhigg C. Hospital clients nurture firm's scheduling software. *Los Angeles Times.* June 21, 2006.

87 *The Interactive Advertising Bureau reported:* Interactive Marketing & Media Fact Pack 2006. Crain Communications Inc. April 17, 2006, http://adage. com/images/random/Interactivefactpack06.pdf.

88 *The Web can also be used to help get you a raise:* Darlin D. Using the Web to get the boss to pay more. *The New York Times.* March 3, 2007.

88 *Recent economic analyses:* Atkinson RD, McKay A. Digital prosperity: Understanding the economic benefits of the information technology revolution. The Information Technology & Innovation Foundation. March 2007, http:// www.itif.org/files/digital_prosperity.pdf.

88 *economists have described two major errors:* Kuhnen CM, Knutson B. The neural basis of financial risk taking. *Neuron* 2005;47:763–70.

88 *uncertainty triggers neural circuits in the amygdala:* Hsu M, Bhatt M, Adolphs R, Tranel D, Camerer CF. Neural systems responding to degrees of uncertainty in human decision-making. *Science* 2005; 310: 1680–3.

89 *The Newspaper Association of America:* NAA releases ABC FAX-FAX analysis, www.naa.org/Global/PressCenter/2006/NAA-RELEASES-ABC-FAS-FAX-ANALYSIS.aspx?lg=naaorg.

89 *In early 2007,* Time *magazine:* Carr D. Threatened by the Internet, Time Magazine slims down. *The New York Times.* January 8, 2007.

89 *In the fall of 2007, seasoned film and TV producers:* Cieply M. New show to begin on MySpace. *The New York Times.* September 13, 2007.

90 *In the United States, this revenue increased to nearly $17 billion:* Online ad revenue sets record for a third year. *The Wall Street Journal.* March 8, 2007.

90 *But in 2006, Rupert Murdoch's News Corporation:* Kehaulani Goo S. Google gambles on Web video. *The Washington Post.* October 10, 2006, http://www.washington-post.com/wp-dyn/content/article/2006/10/09/AR2006100900546.html.

91 *Women are more likely to email friends:* Fallows D. How men and women use the Internet. *Pew Internet & American Life Project.* Washington, DC, 2005, www.pewinternet.org/pdfs/PIP_Women_and_Men_online.pdf.

91 *Dr. Simon Baron-Cohen and colleagues at Cambridge University:* Baron-Cohen S, Knickmeyer RC, Belmonte MK. Sex differences in the brain: Implications for explaining autism. *Science* 2005 4;310:819–23.

91 *Dr. Richard Haier and colleagues at University of California:* Haier RJ, Jung RE, Yeo RA, Head K, Alkire MT. The neuroanatomy of general intelligence: sex matters. *NeuroImage* 2005; 25:320–7.

92 *At the Internet's inception, men initially dominated:* Fallows D. How men and women use the Internet. *Pew Internet & American Life Project.* Washington, DC, 2005.

92 *differences in brain function and structure between the sexes:* The mismeasure of woman. *The Economist.* August 3, 2006, www.economist.com/science/displaystory.cfm?story_id=7245949.

92 *Investigators at the University of Minnesota:* Fulkerson JA, Story M, Mellin A, Leffert N, Neumark-Sztainer D, French SA. Family dinner meal frequency and adolescent development: Relationships with developmental assets and high-risk behaviors. *Journal of Adolescent Health* 2006;39:337–45.

95 *A 2006 Pew Internet study found that nearly 40 percent:* Madden M, Lenhart A. Online dating. *Pew Internet & American Life Project.* March 5, 2006, www.pewinternet.org/pdfs/PIP_Online_Dating.pdf.

96 *The traditional love note:* Zaslow J. Digital love letters are easy to send but hard to cherish. *The Wall Street Journal.* February 9, 2007.

96 *love dramatically changes the brain:* Fisher HE, Aron A, Brown LL. Romantic love: A mammalian brain system for mate choice. *Philosophical Transactions of the Royal Society of London. Series B. Biological Sciences* 2006;361:2173–86.

97 *The prevalence of portable technology:* Hafner K. Laptop slides into bed in love triangle. *The New York Times.* August 24, 2006.

97 *AOL inadvertently posted nineteen million Internet:* Zeller T. Your life as an open book. *The New York Times.* August 12, 2006.

98 *one in four practicing doctors in the United States:* National Center for Health Statistics. Electronic medical record use by office-based physicians: United States, 2005, www.cdc.gov/nchs/products/pubs/pubd/hestats/electronic/electronic.htm.

98 *which would violate the HIPAA (Health Insurance Portability and Accountability Act) regulations:* Stone JH. Communication between physicians and patients in the era of e-medicine. *The New England Journal of Medicine,* 2007;356:2451–4.

98 *Many businesses are now using programs:* Dell K, Cullen LT. Snooping bosses. *Time.* September 11, 2006, http://www.workrights.org/in_the_news/in_the _news_time.html.

98 *using GPS technology to track workers' whereabouts:* Geller A. Bosses keep sharp eye on mobile workers via GPS. *Associated Press.* January 3, 2005, http:// www.workrights.org/in_the_news/in_the_news_associatedpress.html.

98 *Many students haven't a clue:* Finder A. For some, online persona undermines a résumé. *The New York Times.* June 11, 2006.

99 *Ethicist Randy Cohen notes:* Cohen R. Online extracurriculars. *The New York Times Sunday Magazine.* March 11, 2007.

99 *In 2008, an estimated 80 percent of all cell phones:* Sullivan MS. Law May Curb Cellphone Camera Use. PCWorld.com. July 23, 2004, www.pcworld.com/ article/id,117035-page,1/article.html.

99 *The FBI ranks cyber crime:* Bryan-Low C. To catch crooks in cyberspace, FBI goes global. *The Wall Street Journal.* November 21, 2006.

99 *In 2006, there were an estimated five thousand:* Meyer J. Extremists are homing in on the Internet, says Gonzales. *Los Angeles Times.* August 17, 2006.

100 *emotional centers of the brains:* Bloom FE, Beal MF, Kupfer DJ (eds). *The Dana Guide to Brain Health.* The Dana Press. New York, 2003.

100 *viewing gruesome images activates a specific network:* Nitschke JB, Sarinopoulos I, Mackiewicz KL, Schaefer HS, Davidson RJ. Functional neuroanatomy of aversion and its anticipation. *Neuroimage* 2006;29:106–16.

100 *Worldwide cooperation:* Cohen LP. Internet's ubiquity multiplies venues to try web crimes. *The Wall Street Journal.* February 12, 2007.

100 *malicious computer viruses caused an estimated $14 billion:* Bryan-Low C. To catch crooks in cyberspace, FBI goes global. *The Wall Street Journal.* November 21, 2006.

100 *text messaging to report a crime:* Yuan L. Murder, she texted: Wireless messaging used to fight crime. *The Wall Street Journal.* July 2, 2007.

101 *In 2007, an estimated seventy million blogs:* Saranow J. The minutes of our lives. *The Wall Street Journal.* March 2, 2007.

101 *Although blogs are often associated:* Lenhart A, Fox S. Bloggers: A portrait of the internet's new storytellers. *Pew Internet & American Life Project.* Washington, DC, 2006.

101 *Hollywood studios enlist bloggers:* Studios enlist blowers for Oscar campaigns, ruffling feathers in Hollywood hierarchy. *The Wall Street Journal.* February 8, 2007.

102 *In the run-up to the 2008 presidential election in the United States:* Nagourney A. Politics faces sweeping change via the Web. *The New York Times.* April 2, 2006.

Schatz A. Candidates find a new stump in the blogosphere. *The Wall Street Journal.* February 14, 2007.

102   **more than fifty million monthly MySpace users:** Comscore.com. Social networking sites continue to attract record numbers as MySpace.com surpasses 50 million U.S. visitors in May. June 15, 2006, www.comscore.com/press/release.asp?press=906.

102   **YouTube also became an important campaigning force:** Noveck J. YouTube follows the campaign trail. *Los Angeles Times.* July 6, 2007.

102   **Americans reported that they regularly learned about the 2008 presidential campaign:** The Internet's broader role in Campaign 2008. *Pew Internet & American Life Project.* January 11, 2008. http://pewresearch.org/pubs/689/the-internets-broader-role-in-campaign-2008.

102   **The Connecticut research firm PQ Media estimated that $80 million:** PQ Media's Political Media Buying 2008: Preliminary Forecast Analysis. http://www.pqmedia.com/political-media-buying-2008.html.

103   **Professor David Amodio and associates at New York University:** Amodio DM, Jost JT, Master SL, Yee CM. Neurocognitive correlates of liberalism and conservatism. *Nature Neuroscience* 2007;10:1246-7.

103   **research volunteers viewed the faces of presidential candidates:** Kaplan JT, Freedman J, Iacoboni M. Us versus them: political attitudes and party affiliation influence neural response to faces of presidential candidates. *Neuropsychologia* 2007;45:55–64.

CHAPTER 6: BRAIN EVOLUTION: WHERE DO YOU STAND NOW?

105   **we have used self-assessment questionnaires:** Ercoli LM, Siddarth P, Huang S-C, et al. Perceived loss of memory ability and cerebral metabolic decline in persons with the apolipoprotein E-4 genetic risk for Alzheimer's disease. *Archives of General Psychiatry* 2006; 63:442–8.

CHAPTER 7: RECONNECTING FACE TO FACE

116   **many colleges have introduced courses:** Silverstein S. Real world 101: Colleges teach dining, taxes, life. *Los Angeles Times.* June 10, 2006.

116   **everyday social contacts may boost brain power:** Ybarra O, Burnstein E, Winkielmon P, et al. Mental exercising through simple socializing: Social interaction promotes general cognitive functioning. *Personality and Social Psychology Bulletin* 2008; 34:248–59.

117   **it defines our humanity:** Chayer C, Freedman M. Frontal lobe functions. *Current Neurology and Neuroscience Reports* 2001;1:547–5.

118   **One key brain region, the insula:** Blakeslee S. A small part of the brain, and its profound effects. *The New York Times.* February 6, 2007.

118   **Smokers with insular damage:** Naqvi NH, Rudrauf D, Damasio H, Bechara A. Damage to the insula disrupts addiction to cigarette smoking. *Science* 2007:315;531- 4.

118   *The insula partners with other brain centers:* Koenigs M, Young L, Adolphs R, et al. Damage to the prefrontal cortex increases utilitarian moral judgments. *Nature* 2007;446:865–6.
       Carey B. Brain injury said to affect moral choices. *The New York Times.* March 22, 2007.

119   *Dr. John King and his associates at University College in London:* King JA, Blair JR, Mitchell DGX, Dolan RJ, Burgess N. Doing the right thing: A common neural circuit for appropriate violent or compassionate behavior. *NeuroImage* 2006;30:1069–76.

120   *Brain function generally does decline with age:* Williams LM, Brown KJ, Palmer D, et al. The mellow years?: Neural basis of improving emotional stability over age. *Journal of Neuroscience* 2006;26:6422–30.

120   *Dr. Arthur Kramer of the University of Illinois:* Begley S. Parts of brain seem to get better with age. *The Wall Street Journal.* February 17, 2007

121   *a sixteen-year-old driver is twenty times more likely:* The teen driver: Committee on Injury, Violence, and Poison Prevention and Committee on Adolescence. *Pediatrics* 2006;118;2570–81.

121   *In fact, auto accidents:* Centers for Disease Control and Prevention. Web-based Injury Statistics Query and Reporting System (WISQARS) [Online]. (2006). National Center for Injury Prevention and Control, Centers for Disease Control and Prevention (producer), www.cdc.gov/ncipc/wisqars.

121   *Professor Leanne Williams of the University of Sydney:* Williams et al. *Journal of Neuroscience* 2006;26:6422–30.

121   *Other research by Dr. Thomas Hess:* Leclerc CM, Hess TM. Age differences in the bases for social judgments: Tests of a social expertise perspective. *Experimental Aging Research* 2007;33:95–120.
       Begley S. The upside of aging. *The Wall Street Journal.* February 17, 2007.

121   *mature brain is more resilient:* Foreman J. It seems all those birthdays may be making you happy. *Los Angeles Times.* July 16, 2007.

121   *talk therapies can influence brain activation patterns:* Bradley S. Foregrounding language. On the relationship between therapeutic words and the brain. *Psychiatric Annals* 2006;36:289–94.
       Schwartz JM, Stoessel PW, Baxter LR, Martin KM, Phelps ME. Systemic changes in cerebral glucose metabolic rate after successful behavior modification treatment of obsessive-compulsive disorder. *Archives of General Psychiatry* 1996;53:109–13.

122   *Chronic Internet and technology users:* Koezuka N, Koo M, Allison KR, et al. The relationship between sedentary activities and physical inactivity among adolescents: Results from the Canadian community health survey. *Journal of Adolescent Health* 2006;39:515–22.

122   *adopt a healthier lifestyle:* Small et al. *American Journal of Geriatric Psychiatry* 2006;14:538–45.

124 ***Keep your brain fit with off-line mental aerobics:*** Small G, Vorgan G. *The Longevity Bible*. Hyperion, New York, 2006.
Begley S. How to keep your aging brain fit: Aerobics. *The Wall Street Journal*. November 16, 2006.

124 ***Functional MRI studies at UCLA:*** Wang AT, Lee SS, Sigman M, Dapretto M. Reading affect in the face and voice. Neural correlates of interpreting communicative intent in children and adolescents with autism spectrum disorders. *Archives of General Psychiatry* 2007;64:698–708.

127 ***Dr. Adam Joinson of the Institute of Educational Technology:*** Joinson AN. Self-esteem, interpersonal risk, and preference for e-mail to face-to-face communication. *Cyperpsychology & Behavior* 2004;7:472–8.

131 ***volunteers with low self-esteem were significantly more likely to choose email:*** Joinson AN. *Cyperpsychology & Behavior* 2004;7:472–8.

131 ***Low self-worth also can lead to the online disinhibition effect:*** Suler J. The online *disinhibition* effect. *CyberPsychology & Behavior* 2004;7:321–6.

131 ***become cyber bullies:*** Harmon A. Internet gives teenage bullies weapons to wound from afar. *The New York Times*. August 26, 2004.

131 ***brain regions that normally inhibit such aggression:*** New AS, Hazlett EA, Buchsbaum MS, et al. Blunted prefrontal cortical 18fluorodeoxyglucose positron emission tomography response to meta-chlorophenylpiperazine in impulsive aggression. *Arch Gen Psychiatry* 2002;59:621–9.

132 ***pinpointed the brain regions that control optimism:*** Sharot T, Riccardi AM, Raio CM, Phelps EA. Neural mechanisms mediating optimism bias. *Nature* 2007: 450: 102–5.

134 ***volunteers engaged in an empathic exercise:*** Carr L, Iacoboni M, Dubeau M-C, Mazziotta JC, Lenz GL. Neural mechanisms of empathy in humans: A relay from neural systems for imitation to limbic areas. *Proceedings of the National Academy of Science USA* 2003;100:5497–502.

134 ***Dr. Tania Singer and colleagues at the Institute of Neurology:*** Singer T, Seymour B, O'Doherty J, Kaube H, Dolan RJ, Frith CD. Empathy for pain involves the affective but not sensory components of pain. *Science* 2004;303:1157–62.

137 ***Technology writer Danny O'Brien polled his list:*** Thompson C. Meet the life hackers. *The New York Times*. October 16, 2005.

142 ***Dr. Richard Davidson's team at the University of Wisconsin:*** Davidson RJ, Kabat-Zinn J, Schumacher J, et al. Alterations in brain and immune function produced by mindfulness meditation. *Psychosomatic Medicine* 2003;65:564–70.
Geirland J. Buddha on the brain: The hot new frontier of neuroscience: Meditation! (just ask the Dalai Lama.) *Wired*. February 2006, www.wired.com/wired/archive/14.02/dalai.html.

143 ***and possibly even your life expectancy:*** Small G, Vorgan G. *The Longevity Bible*. Hyperion, New York, 2006.

145 ***the left hemisphere controls writing through Broca's area:*** Wing AM. Motor

control: Mechanisms of motor equivalence in handwriting. *Current Biology* 2000;10:R245–8.

CHAPTER 8: THE TECHNOLOGY TOOLKIT

150 *once again our brains look for familiar brands, like Apple or IBM:* Born C, Meindl T, Poeppel E, et al. Brand perception: Evaluation of cortical activation using fMRI. *Annual Meeting of the Radiological Society of North America*, November 28, 2006.

151 *Caltech investigator Ming Hsu and his colleagues used functional MRI:* Hsu M, Bhatt M, Adolphs R, Tranel D, Camerer CF. Neural systems responding to degrees of uncertainty in human decision-making. *Science* 2005; 310: 1680–3.

154 *As Will Schwalbe and David Shipley wrote in their book:* Shipley D, Schwalbe W. *Send.* Alfred A. Knopf, New York, NY, 2007.

158 *people are inefficient in the way they search on the Web:* Barbara M, Zeller T. A Face Is Exposed for AOL Searcher No. 4417749. *The New York Times.* August 9, 2006.

161 *Washington, New York, and California:* Richtel M. Hands on the wheel, not on the BlackBerry keys. *The New York Times.* May 12, 2007.
      Cooper C. Legislators aim at a new misdeed on the road: DWT. *The Wall Street Journal.* March 14, 2007.

162 *A recent National Highway Traffic Safety Administration Study:* Insurance Information Institute, Inc. Cell Phones and Driving, http://www.iii.org/media/hottopics/insurance/cellphones/.

162 *comparable to those associated with driving while drunk:* Strayer DL, Drews FA, Crouch DJ. A comparison of the cell phone driver and the drunk driver. *Human Factors* 2006;48:381–91.

162 *Cell phone activity like making a call or text messaging, three brain regions:* Osaka M, Komori M, Morishita M, Osaka N. Neural bases of focusing attention in working memory: an FMRI study based on group differences. *Cognition Affective Behavioral Neuroscience* 2007;7:130–9.

162 *Peripheral vision is particularly affected:* Wood J, Chaparro A, Hickson L, et al. The effect of auditory and visual distracters on the useful field of view: Implications for the driving task. *Investigative Ophthalmology and Visual Science.* 2006;47:4646–50.

167 *TV channels that have found a niche on the Web:* White B. TV channels move to the Web, think outside the cable box. *The Wall Street Journal.* August 20, 2007.

168 *perception of control is associated with subcortical dopamine:* Declerck CH, Boone C, De Brabander B. On feeling in control: a biological theory for individual differences in control perception. *Brain and Cognition* 2006;62:143–76.

168 *To understand how our brains respond when we face a fear:* Mobbs D, Petrovic P, Marchant JL, et al. When fear is near: Threat imminence elicits prefrontal-periaqueductal gray shifts in humans. *Science* 2007;317:1079–83.

169 *the U.S. Department of Veterans Affairs reported:* Secretary Nicholson provides

update on stolen data incident: Data matching with Department of Defense providing new details. U.S. Department of Veterans Affairs, http://www1.va.gov/opa/pressrel/pressrelease.cfm?id=1134.

169 *Automated PC hackers trying to crack a password:* University of Massachusetts Lowell, IT department, www.uml.edu/it/default.html.

170 *Pornography is readily accessible on the Web, and many parents express concern:* Zetter K. How best to protect kids from online porn. *San Francisco Chronicle.* February 12, 2006, http://www.sfgate.com/cgi-bin/article.cgi?f=/c/a/2006/02/12/FILTERING.TMP

171 *another new twist to online privacy:* Pham A, Menn J. Google Maps redraw the realm of privacy. *Los Angeles Times.* August 7, 2007.

172 *Food and Drug Administration (FDA) is considering approval of a surgically implanted device:* Burton TM, Wilde Mathews A. Monitoring your heart via the Internet. *The Wall Street Journal.* February 28, 2007.

173 *eight out of ten American Internet users have searched for health information:* Fox S. Online health search 2006. *Pew Internet & American Life Project.* October 29, 2006, www.pewinternet.org/pdfs/PIP_Online_Health_2006.pdf.

173 *To increase the likelihood of obtaining valid answers to health questions:* Kushner D. Well connected: Finding trustworthy medical info online. *AARP* March & April 2007; pp. 40–42.

174 *Brain imaging studies of hypochondriacs:* van den Heuvel OA, Veltman DJ, Groenewegen HJ, et al. Disorder-specific neuroanatomical correlates of attentional bias in obsessive-compulsive disorder, panic disorder, and hypochondriasis. *Archives of General Psychiatry* 2005;62:922–33.
Hakala M, Vahlberg T, Niemi PM, Karlsson H. Brain glucose metabolism and temperament in relation to severe somatization. *Psychiatry Clinical Neurosciences* 2006;60:669–75.

175 *Web-based resources are now available:* Francis T. How to find a good doctor, www.inod.org/news.html.

176 *many doctors remain concerned:* Brooks RG, Menachemi N. Physicians' use of email with patients: Factors influencing electronic communication and adherence to best practices. *Journal of Medical Internet Research* 2007;8(1):e2, www.jmir.org/2006/1/e2.

176 *guidelines to help physicians use email:* Robertson J. Guidelines for physician-patient electronic communications. American Medical Association, www.ama-assn.org/ama/pub/category/2386.html.

178 *mental stimulation may improve memory and brain health:* Greene K. Putting brain exercises to the test. *The Wall Street Journal.* February 3, 2007.

178 *Brain scanning studies using PET:* Small GW, Silverman DHS, Siddarth P, et al. Effects of a 14-day healthy longevity lifestyle program on cognition and brain function. *American Journal of Geriatric Psychiatry* 2006;14:538–45.
Haier RJ, Siegel BV, MacLachlan A, et al. Regional glucose metabolic changes

after learning a complex visuospatial/motor task: A positron emission to-mographic study. *Brain Research* 1992;570:134–43.

180   ***Recent research suggests that stimulating our brains:*** Small GW, Silverman DHS, Siddarth P, et al. Effects of a 14-day healthy longevity lifestyle program on cognition and brain function. *American Journal of Geriatric Psychiatry* 2006;14:538–45.

180   ***efficiency of the neural circuitry:*** Alter A. Is this man cheating on his wife? *The Wall Street Journal.* August 10, 2007.

Bailenson JN, Yee N. *Digital chameleons:* Automatic assimilation of nonverbal gestures in immersive virtual environments. *Psychological Science* 2005;16:814–9.

CHAPTER 9: BRIDGING THE BRAIN GAP: TECHNOLOGY AND THE FUTURE BRAIN

181   ***within the category of Digital Natives there are two subgroups:*** Howe N, Strauss W. *Millennials Rising: The Next Great Generation.* Vintage, New York, NY. 2000.

Lancaster LC, Stillman S. *When Generations Collide: Who They Are. Why They Clash. How to Solve the Generational Puzzle at Work.* Collins, New York, NY. 2003.

182   ***Journalist Carol Hymowitz has described how:*** Hymowitz C. Managers find ways to get generations to close culture gaps. *The Wall Street Journal.* July 9, 2007.

187   ***Researchers have already developed a neurochip:*** Than K. Brain cells fused with computer chip. *LiveScience.* March 27, 2006, www.livescience.com/health/060327_neuro_chips.html.

187   ***epileptic patients to control a computer cursor:*** Carey B. Brain power: Mind control of external devices. *LiveScience.* March 17, 2005, www.livescience.com/health/050317_brain_interface.html.

Schwartz AB, Cui XT, Weber DJ, Moran DW. Brain-controlled interfaces: movement restoration with neural prosthetics. *Neuron* 2006;52:205–20.

187   ***Brain-computer interface technologies detect and translate:*** Coyle S, Ward T, Markham C, McDarby G. On the suitability of near-infrared (NIR) systems for next-generation brain-computer interfaces. *Physiological Measurement* 2004;25:815–22.

187   ***hook up a human volunteer to mentally type:*** Taylor C. Surfing the Web with nothing but brainwaves. CNN Money.com. July 24, 2006, http://money.cnn.com/2006/07/21/technology/googlebrain0721.biz2/index.htm.

Karim AA, Hinterberger T, Richter J, et al. Neural internet: Web surfing with brain potentials for the completely paralyzed. *Neurorehabilitation and Neural Repair* 2006;20:508–15.

188   ***While volunteers were hooked up to this skin-reading technology:*** Peplow M. Mental ping-pong could aid paraplegics. *Nature News.* August 27, 2004, http://www.bioedonline.org/news/news.cfm?art=1126.

189   ***a new apparatus that uses a photosensitive protein:*** Chen I. The beam of light that flips a switch that turns on the brain. *The New York Times.* August 14, 2007.

INDEX